C0-AYJ-714

Achieving Justice in Genomic Translation

Achieving Justice in Genomic Translation

Rethinking the Pathway to Benefit

Edited by

WYLIE BURKE, KELLY EDWARDS, SARA GOERING, SUZANNE HOLLAND, AND SUSAN BROWN TRINIDAD

Oxford University Press, Inc., publishes works that further
Oxford University's objective of excellence
in research, scholarship, and education.

Oxford New York
Auckland Cape Town Dar es Salaam Hong Kong Karachi
Kuala Lumpur Madrid Melbourne Mexico City Nairobi
New Delhi Shanghai Taipei Toronto

With offices in
Argentina Austria Brazil Chile Czech Republic France Greece
Guatemala Hungary Italy Japan Poland Portugal Singapore
South Korea Switzerland Thailand Turkey Ukraine Vietnam

Published by Oxford University Press, Inc.
198 Madison Avenue, New York, New York 10016

www.oup.com

Library of Congress Cataloging-in-Publication Data

Achieving justice in genomic translation : re-thinking the pathway to
benefit / edited by Wylie Burke ... [et al.].
p. ; cm.
Includes bibliographical references and index.
ISBN 978-0-19-539038-4
1. Human genetics—Research—Moral and ethical aspects.
2. Genomics—Research—Moral and ethical aspects.
3. Genetic screening—Moral and ethical aspects. I. Burke, Wylie.
[DNLM: 1. Genetic Research—ethics. 2. Bioethical Issues.
3. Genetic Predisposition to Disease.
4. Genetic Testing—ethics. 5. Translational Research—ethics. WB 60]
QH438.7.A26 2011
174.2'96042—dc22
2011011621

9 8 7 6 5 4 3 2 1
Printed in the United States of America
on acid-free paper

ACKNOWLEDGMENTS

This book is the product of several years of collaborative research and dialogue within our Center for Genomics and Healthcare Equality, a Center for Excellence in Ethical, Legal, and Social Implications Research funded by the National Human Genome Research Institute (Grant P50 HG03374). The Testing Justice Project, funded by The Greenwall Foundation, provided additional support for Drs. Edwards, Goering, and Holland and gave us the opportunity to truly explore and develop the normative core of the book. We would also like to acknowledge the numerous researchers, community partners, trainees, and research staff who have connected with us and the Center over the past five years. The experiences and insights you have shared with us have indelibly shaped our thinking and core commitments to the responsible conduct of research.

CONTENTS

CONTRIBUTORS

Wylie Burke, MD, PhD
Department of Bioethics and Humanities
Center for Genomics and Healthcare Equality
University of Washington
Seattle, WA

Patricia Deverka, MD, MS
Center for Medical Technology Policy
Baltimore, MD

Kelly Edwards, PhD
Department of Bioethics and Humanities
Center for Genomics and Healthcare Equality
University of Washington School of Medicine
Seattle, WA

Stephanie Malia Fullerton, DPhil
Department of Bioethics and Humanities
Center for Genomics and Healthcare Equality
University of Washington
Seattle, WA

Sara Goering, PhD
Department of Philosophy
Program on Values in Society
University of Washington
Seattle, WA

Nora B. Henrikson, PhD, MPH
Group Health Research Institute
Group Health Cooperative
Seattle, WA

Suzanne Holland, PhD
Department of Religion
University of Puget Sound
Tacoma, WA

Rosalina James, PhD
Department of Bioethics and Humanities
Center for Genomics and Healthcare Equality
University of Washington
Seattle, WA

Maureen Kelley, PhD
Treuman Katz Center for Pediatric Bioethics
Seattle Children's Research Institute
Department of Pediatrics
University of Washington School of Medicine
Seattle, WA

Patricia C. Kuszler, MD, JD
School of Law
Center for Law, Science and Global Health
Department of Bioethics, School of Medicine
Department of Health Services, School of Public Health
University of Washington
Seattle, WA

Anne-Marie Laberge, MD, PhD
Medical Genetics Division, Centre Hospitalier Universitaire Sainte-Justine
Department of Pediatrics
Université de Montréal
Montréal, Québec, Canada

Martine D. Lappé, PhD
Department of Social and Behavioral Sciences
University of California
San Francisco, CA

Mitzi L. Murray, MD
Treuman Katz Center for Pediatric Bioethics
Seattle Children's Research Institute
Department of Pediatrics
University of Washington
Seattle, WA

Nancy Press, PhD
School of Nursing and Department of Public Health & Preventive Medicine, School of Medicine
Oregon Health and Science University
Portland, OR

Catharine Riley, MPH, PhD candidate
Institute for Public Health Genetics
School of Public Health
University of Washington
Seattle, WA

Helene Starks, PhD, MPH
Department of Bioethics and Humanities
Center for Genomics and Healthcare Equality
University of Washington
Seattle, WA

Holly K. Tabor, PhD
Treuman Katz Center for Pediatric Bioethics
Seattle Children's Research Institute
Department of Pediatrics
University of Washington School of Medicine
Seattle, WA

Susan Brown Trinidad, MA
Department of Bioethics and Humanities
University of Washington
Seattle, WA

David L. Veenstra, PharmD, PhD
Pharmaceutical Outcomes Research and Policy Program (PORPP), and Institute for Public Health Genetics
Center for Genomics and Healthcare Equality
University of Washington
Seattle, WA

Carolyn (Cindy) A. Watts, PhD
Arthur Graham Glasgow Professor and Chair
Department of Health Administration
Virginia Commonwealth University
Richmond, VA

Benjamin S. Wilfond, MD
Treuman Katz Center for Pediatric Bioethics
Seattle Children's Research Institute
Department of Pediatrics
University of Washington School of Medicine
Seattle, WA

Achieving Justice in Genomic Translation

1

Making Good on the Promise of Genetics: Justice in Translational Science

SARA GOERING, SUZANNE HOLLAND, AND KELLY EDWARDS

The promise of biomedical research depends, quite pragmatically, on the translation of basic scientific findings into therapeutic applications that improve health. The translational imperative is especially strong for areas of study that have involved significant public investment and carry a substantial expectation of benefit, such as human genetics research. Yet translation is often an unmet goal: according to one general review, only 5% of "highly promising" basic science findings ever lead to the development of applications suitable for clinical use, and only 1% are used according to clinical guidelines (Contopoulos-Ioannidis, Ntzani, and Ioannidis 2003).

Science policy makers are now paying increased attention to translational goals. In 2006, the National Institutes of Health launched the Clinical and Translational Science Award program, which gives multimillion-dollar grants to interdisciplinary research teams of researchers, with the aims of moving new knowledge gained through bench science to the bedside and into regular clinical practice, and ultimately to improve community health. Yet while efficiency of translation is no doubt important, more fundamental questions remain about justice and utility. Who benefits, and what counts as a benefit?

"Who benefits?" calls for attention to the distribution of the benefits and burdens of research, and to the values that do, and should, drive research. "What counts as a benefit?" is a question about who gets a say in how research is done and how its products are evaluated. Because our focus in this book is justice in health care, we are particularly interested in whether biomedical genetics research can do anything to address existing health disparities, and how such research could be transformed to take seriously the needs of the marginalized and medically underserved. A fundamental premise of this book, therefore, is that we ought to expose, reexamine, and transform the values currently driving translational science in service of more just and effective ends.

The questions of what counts as a benefit and who benefits clearly intersect in this arena. For instance, if basic genomic science identifies a genetic marker that is associated with an increased risk of developing type 2 diabetes, a dedication to speedy translation might recommend that a genetic test for that marker be developed and offered to patients, either directly or in the clinic. But how beneficial is such a test? And who benefits from such information? As a society with a predilection for technology, we tend to surge forward, eager to have more knowledge, without necessarily taking the time to assess the relative merits of that information. In this case, it is not clear that individuals who appear to have a greater genetic risk will necessarily alter their behaviors (e.g., improve diet or increase exercise) to promote better health. It is also possible that people who the test indicates do not have the marker might consider themselves genetically lucky and therefore feel less motivated to eat right and exercise. Furthermore, marginalized and medically underserved people may see the provision of such a test as relatively useless, particularly if it comes at the expense of other basic care they lack and if acting on the resulting risk-reduction recommendations is difficult or impossible in their circumstances. If the ultimate aim of translational genomics is to improve health outcomes, it might make more sense to try to address the problem of widespread obesity due to lifestyle habits rather than focus on genetic susceptibility testing (Burke et al. 2008). Other examples—genetic markers for psychiatric illness, for Alzheimer's disease, or for cancer—are readily available, and concerns about who benefits and who might be harmed are even sharper when the conditions in question are associated with significant stigma, discrimination, and exclusion.

Many advocates for speedier genomic translation presume that advances in our understanding of genetics related to common diseases (heart disease, diabetes, etc.) will help to resolve existing racial/ethnic health disparities. Federal funding priorities reflect this connection as well, such as the 2008 NIH Intramural Center for Genomics and Health Disparities, which became the trans-NIH Center for Research on Genomics and Global Health. Our concern is that without transformation of standard research practices, existing disparities may in fact be exacerbated by the push for translation of genetic research. Our project here is to examine the ways in which health disparities can be narrowed rather than widened through attention to the process and outcomes of genetic research. This book is a product of collaboration among an interdisciplinary team of investigators at the Center for Genomics and Healthcare Equality at the University of Washington, a Center of Excellence funded by the National Human Genome Research Institute.

Throughout this volume, we employ a normative framework intended to challenge and strengthen the translational pathway at each step, from the formulation of research questions to evaluation of outcomes. The framework—which we call "responsive justice"—focuses on the questions of what counts as a benefit; who determines which benefits and outcomes warrant pursuing; and what responsibilities researchers, and the research enterprise as a whole, may have to assure a more equitable distribution of research benefits (Goering, Holland, and Fryer-Edwards 2008). We argue that translation of genetic research can be both more effective and more just when it incorporates not only the interests of researchers, industry, and funders but, crucially, the interests of communities affected by the research, particularly when the communities represent traditionally medically underserved and non-majority groups.

The book presents a series of case studies of genetic innovation across the translational pathway, from inception to outcomes. Each chapter foregrounds questions of health benefit and justice, with the goal of making more visible the implicit, value-based assumptions present in all research. By explicitly calling attention to questions of benefit and justice at each phase of research, and by providing recommendations for more just practices, we hope to spotlight opportunities for funders, policymakers, researchers, and the general public to make more intentional choices about how translational research is done, and thus to improve the likelihood that innovations can lead to improvements in community health.

In this introductory chapter, we first lay out our conception of the translational pathway. We then outline the responsive justice framework in greater detail, showing how it comes into play across the translational pathway. Finally, we provide a brief preview of the chapters that follow.

THE TRANSLATIONAL PATHWAY

Over the past decade, genomic research has identified a number of linkages between genetic markers and common diseases such as diabetes, heart disease, cancer, depression, and Alzheimer's disease. Translating this new knowledge into useful treatments or risk-reduction strategies for individuals and for improved public health has been more difficult, given the complexity of gene–environment interactions as well as the lack of coordination among researchers in discovery science, clinical medicine, and public health. In analyzing the process of moving new knowledge from bench to bedside and beyond, researchers have come to speak of a "translational pathway" and have outlined various models of how the process works. In its earliest formulation, the translational pathway was conceptualized as having two main transition points: getting from the laboratory bench to human testing, and from clinical study results to daily clinical practice (Woolf 2008). Later models elaborated upon this view. For example, Westfall and colleagues (2007) suggested the pathway includes three transitions: moving basic genomic discoveries into candidate health applications such as genetic tests; assessing the value of the test or intervention and creating evidence-based guidelines for clinical practice; and disseminating those guidelines and integrating them into clinical practice. Following on this, Khoury and colleagues (2007) proposed the addition of a fourth phase of translation, which evaluates the "real world" health benefits of a genetic intervention once it has been integrated into clinical practice.

This fourth phase focuses on determining how or whether various implementation practices are linked with public health outcomes at the population level. So, for instance, if guidelines recommended that state health departments expand the range of genetic tests included in newborn screening programs, outcomes researchers would seek to ascertain whether expanded screening did, in fact, improve health outcomes. Most translational research and funding focuses at the earliest stages of translation, with scant attention to implementation and health impact research (Woolf 2008). Yet prioritizing these earlier stages of translational research may hasten to market genetic tests that have limited clinical utility and ignore the complex causal nexus of many genetically linked diseases. Paradoxically, an overemphasis on improving the efficiency of translation may actually fail to benefit individuals and families, particular groups, or even public health generally. In the worst-case scenarios, such investments could bring harm to the very populations most in need of benefit—populations already suffering from health disparities—if not directly, then through opportunity costs of distracting funding and resources to these nonbeneficial technologies.

A New Model of the Translational Pathway

Our vision of the translational pathway (Starks et al. forthcoming) builds on prior models. It includes four main stages of research which we call *discovery*, *development*, *delivery*, and *outcomes* (see Figure 1-1). We highlight the importance of handoffs and transitions between these phases of research for effective translational science. Our model of the translational pathway is distinct from its predecessors in two main respects (see Table 1-1). First, we propose the explicit inclusion of a critical reflective process, a step in which researchers pause to reassess their work and evaluate its effectiveness and possibility for just impacts. This critical step is what we call "*assessment and priority setting*." It highlights the important decisions that are made in conceptualizing which research projects to pursue, which methods to use, and which populations or diseases to target. Second, our conception of translation follows an iterative cycle rather than a linear progression. In our model, effective translation requires a return to assessment and priority setting between each phase of research, and we recognize significant overlap between purportedly distinct phases of research.

Assessment and priority setting indicates the need for attending not only to the assumptions of particular research programs but also to the socio-political-environmental *context* in which research takes place. Literature on health disparities in the United States demonstrates significant health differences between racial/ethnic groups even when access to care is held constant (Smedley et al. 2003).

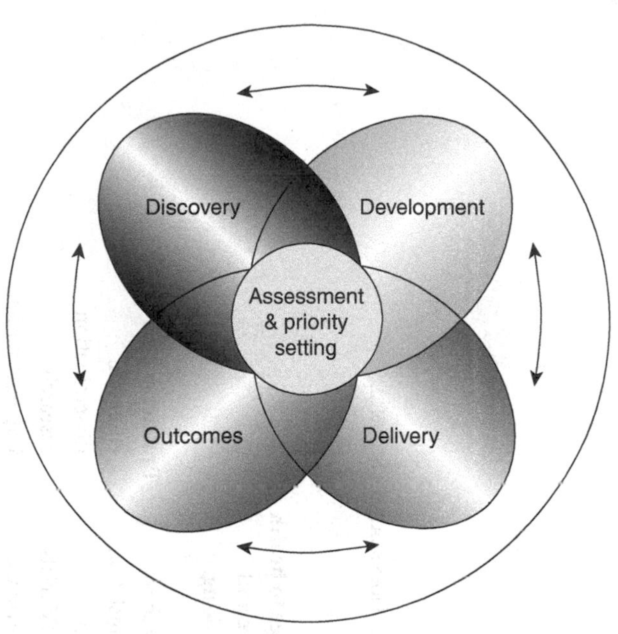

Figure 1-1 Translational Cycle, Center for genomics and healthcare equality University of Washington.

Studies designed to investigate the reasons underlying such disparities might focus on access to care; quality of care received; environmental, behavioral, or nutritional factors affecting disease burden; or possible genetic susceptibility groupings. The process of figuring out which health problems deserve researcher attention, and what kinds of studies should be funded, is the work specific to assessment and priority setting. As the authors explore in the chapters dealing with Alaska Native youth suicide (James and Starks, Chapter 11), and the effects of privatization and market-incentives on research (Kuszler, Chapter 2), a number of factors play into decisions about which research gets taken up and pursued. Foregrounding some of these contextual features—incentive structures, regulatory inconsistencies, funding streams, the relevance of social determinants of health—can bring into sharper relief the hurdles to making headway in relation to health disparities. This book makes the case that explicit and repeated attention to assessment and priority setting reveals opportunities for policy and practice development that may shape research in ways that are more likely to have a broader and more just public health impact.

The cyclical and iterative nature of our model acknowledges the reality that translational science does not simply proceed straightforwardly along a set course, nor should it. Problems identified at any stage might require a return to earlier stages, either to revisit former conclusions or to reassess the relative merits of moving forward on the pathway. Our assessment and priority setting step—the central point of our model—encourages this kind of looping effect. At a workshop

Table 1-1. Evolving Definitions of the Stages of Translation Science

Authors	T_0	T_1	T_2	T_3	T_4
Sung et al., 2003 (*3*)		**Basic research to human studies**	**New knowledge to health care practice and decision-making**		
Westfall et al., 2007 (*5*); Dougherty et al., 2008 (*4*); Rubio et al., 2010 (*15*)		**Basic research to human clinical research** in preclinical & animal trials: translation to human populations in case studies, Phase I & II trials	**Clinical research in humans**: translation to practice in Phase III & IV trials; systematic reviews; guideline development	**Practice-based research**: dissemination research; implementation research; quality evaluation; cost-effectiveness	

Khoury et al., 2007 (*7*)		**Discovery to candidate health application**: Phase I & II clinical trials; observational studies	**Health application to evidence-based practice guidelines**: Phase III clinical trials; observational studies; evidence synthesis and guidelines development	**Practice guidelines to health practice**: dissemination research; implementation research; diffusion research; Phase IV clinical trials	**Practice to population health impact**: outcomes research; population monitoring of morbidity, mortality, benefits, and risks
Starks et al., Forthcoming	**Assessment & priority setting**: What is the state of the evidence? Who is involved with setting priorities and choosing what research is done?	**Discovery to development**: what are the mechanisms of health and disease that can be identified to help develop interventions and products?	**Development to delivery**: what steps are needed to transition basic science mechanisms to interventions and products with potential for health applications?	**Delivery to outcomes**: what is needed to adopt and implement new health interventions and products in practice?	**Outcomes to new discovery**: what outcomes result (both health and non-health, i.e., products, markets, etc.), and how should they inform future research?

we attended, a cancer vaccine researcher related a story that illustrates how this plays out on the ground. She was standing in the hallway, sharing with a colleague her excitement about the fact that she had finally been able to isolate a cancer vaccine (a discovery-phase enterprise) when a question from a passing colleague stopped her in her tracks. The second colleague, who had worked in Africa, asked whether she had considered how the vaccine could be delivered to remote communities of people who needed it (a delivery-phase concern). The researcher told us that, until then, she had not considered how matters such as cost, delivery obstacles, and infrastructure requirements could influence the ultimate usefulness of her discovery. Recognizing that her goal was for her cancer vaccine to reach a broader population, she jettisoned her initial approach and returned to the lab to look for a simpler and more easily supported vaccine model that could be more readily scaled-up for the communities most in need.

The researcher's chance meeting with a colleague reoriented her perspective on the discovery and compelled her to return to the laboratory, with her eye on improving health outcomes for the medically underserved. Our model aims to routinize this kind of reflection and redirection, so as to rely less on such serendipitous encounters. Given the public nature of funding for much genetic innovation, and the promissory claims of its developers, reflective assessment work at every stage is vital to ensure that translational science benefits public health. In this book we employ a justice-based perspective to make the case that researchers and funders should pay particular attention to the needs of vulnerable and historically medically underserved groups. On our formulation of the translational pathway, movement to the next phase of the translational cycle is always provisional; reflections in the assessment and priority setting step may recommend a return to an earlier phase of research, or may advocate a push forward. As such, our model reflects the iterative nature of translational science. Though this process may, in some instances, slow the progress along the pathway, we argue that ensuring justice at each stage of the pathway warrants this delay.

RESPONSIVE JUSTICE

While every grant writer can make an argument for how his or her line of research stands to benefit humanity, a review of benefit—and concerns about justice—at each stage in the pathway requires more than rhetorical attention. In this book, we focus on responsive justice, which includes three features: redistribution, recognition, and responsibility. *Distributive justice* primarily has to do with issues of access to, and allocation of, benefits and burdens—who gets what, and on the basis of what understanding of fairness. It is the aspect of justice most familiar to public health researchers and the general public, particularly given recent national debates over the provision of health insurance. *Recognition* highlights the importance of understanding the views and values of under-represented groups, especially those that have faced significant discrimination in the past, and establishes a process that allows them to participate in decision making. *Responsibility* emphasizes the obligation of those with more power in identifying injustice and in making sure that fair distribution and recognition are achieved. In our view, these three elements of justice—which collectively comprise responsive justice—are important to consider at each stage of the translational pathway. In what follows, we describe the elements more fully and offer brief examples to show how the translational pathway might be transformed by careful reflection about justice and utility, and by a commitment to the responsive justice framework.

Distributive Justice

Fair distribution is a central concern for theories of justice. When we think of *distributive justice*, we ask ourselves *who should get what and how benefits will be allocated.* Are there particular patterns of distribution that are just (e.g., equal shares for all), or is justice tied to equality of opportunity, or to entitlement, to name just a few options? In the translational pathway for genome science, concerns about distribution of benefits might mean attending to how much research money is set aside to address particular problems (an assessment and priority setting problem); for example, should "orphan" diseases receive significant public funding, even though they affect relatively few individuals? Distributive justice asks whether new genetic tests are priced beyond the reach of some groups, or whether genetic tests or counseling are available to people who do not live near large cities.

An example from the development phase of the pathway for genetic research—the issue of inclusion of women and minorities in clinical trials—will serve to illuminate how distributive considerations can affect scientific practice. The results of such studies are used to guide the development of clinical practice standards, but if trials include only a narrow sample of the general population, approved drugs and/or therapies may fail to work as predicted in segments of the population that differ from the subject pool (e.g., genetically or in terms of diet, behavior, environmental exposures, etc.). For instance, middle-aged white men were long considered the ideal subjects for clinical trials, because they are relatively easy to recruit and do not experience the monthly hormonal cycles of women. Women's hormonal changes were considered problematic for the collection of controlled data, and in the interest of protecting fetal health researchers also treated all women as potentially pregnant, and therefore particularly vulnerable as subjects. Data from these studies that used only men were then used to develop general recommendations for the use of medications and procedures for patients, including women and minorities. There are two problems entailed by such an approach. First, we now know that some treatments work differently for different kinds of patients; that is, the fact that a drug is safe and effective for one narrowly defined group may not necessarily mean it is safe for all. Second, while the safety of study drugs for middle-aged white males may be demonstrated by such trials, the real trial for other kinds of people occurs in clinical practice—that is, outside the protective context of close observation and control of a clinical trial.

Acknowledgement of this potentially problematic generalization (as well as concern about access to the direct benefits of participation in clinical trials) resulted in changes to NIH policy, which now requires the recruitment and inclusion of women and minorities unless "a clear and compelling rationale and justification establishes to the satisfaction of the relevant Institute/Center Director that inclusion is inappropriate with respect to the health of the subjects or the purpose of the research" (NIH 2001). Notably, cost is not a sufficient reason for exclusion. Thus, groups that had previously been considered either difficult to recruit or particularly vulnerable, and therefore deserving of special (potentially expensive) protections in research, have been more carefully and systematically included in clinical trials. Indeed, similar arguments are now made for the inclusion of pregnant women and children, populations once considered too vulnerable for participation in clinical trials.

We believe this is a step forward, and without a doubt, fair access and fair allocation of benefits and burdens serve as an important component of justice. Too often, though, these kinds of distributive concerns stand in for the whole of justice. That is, theorists commonly envision distributional patterns of fairness that apply to all based on an assumption of similarity. In the above example, the assumption is that if participation in clinical trials provides a benefit, then it should be available to all. Researchers should therefore be pressured to recruit purposefully to ensure the representation of women and minority groups, rather than simply employing a policy that states that anyone may participate and then enrolling the usual suspects. What this emphasis on fair distribution misses is that some of the reasons behind a lack of participation are not mere problems of access. There are historical reasons why certain minority groups, for example American Indians or African Americans, might distrust the mainstream majority. In the past, there simply have not been trustworthy research or contractual practices between minority and majority groups (e.g., the 40-year-long, federally funded Public Health Service study of untreated syphilis in poor African American men in Tuskegee Alabama; Gamble 2006; Reverby 2009). Therefore, individuals in minority communities may have a deep sense of distrust of research and fear a lack of control over what is done with their genetic material despite assurances from the research team. Some individuals may find the focus of genomic research problematic, insofar as it tends to emphasize inherent genetic components to the exclusion of environmental factors. Others may be concerned that any drug or therapy that comes out of the clinical trial will be expensive and available only to those with good insurance, and therefore unlikely to help people in their communities. People from historically marginalized groups may worry about being used for the benefit of others; even if they might get some minor benefit out of participation, they may feel exploited, given the relative imbalance of benefits gained. All of these concerns are understandable, and yet these perspectives are too often neglected.

Responsive justice insists on starting with the real-world needs of socially situated groups that experience systematic disadvantage. Our framework calls for a *response* to their situation, one that requires recognizing their needs from the outset.

Recognition

To better identify and take account of the concerns of historically disenfranchised minority groups, responsive justice also emphasizes the importance of *recognition.* Recognition indicates *a fundamental awareness of and respect for the person or communities with whom research is being conducted*, and to whom (one hopes) the clinical benefits of research will be returned. In its simplest form, recognition can occur when a genetic researcher steps out of the lab to visit an affected community or meet an affected family. This meeting, and the researcher's recognition of what the life circumstances of the other might be like, can provide renewed motivation and commitment to finding an insight that might help.

A research system that rewards efficiency, prioritizes product development, and maximizes profit is one in which researchers' careers are made and sometimes broken, but unfortunately it is not one that encourages paying attention to the values, needs, and power inequalities of communities within which research is being conducted. The present system of research and health care delivery, we submit, more often than not fosters misrecognition. Whenever individuals or groups are fully ignored, unfairly (if unintentionally) stereotyped, categorized as inferior or irrelevant, denied participation in decision making, or have their participation undervalued or marginalized, they are "misrecognized."

Ensuring recognition along the translational pathway requires more than simply listening to minority voices and offering them attention and respect. Rather, it will demand a fairly radical restructuring of the decision-making and power structures of the traditional research system. Following Nancy Fraser, we argue that recognition requires "participatory parity" (Fraser and Honneth 2003). Under conditions of participatory parity, groups that have been misrecognized must be provided the opportunity to voice their claims of misrecognition and to partner in the crafting of solutions. These are more than concessions and compromises; they are actual claims on the dominant group to consider incorporating alternative worldviews and values in guiding the translational process.

From a responsive justice perspective, in order to address health disparities of the medically underserved and marginalized who are misrecognized in our current system of health care and research, those individuals need to be involved in the discussion of what kinds of research are undertaken and in what ways, and that needs to happen in the earliest phases of research. Under the current model of translation (Khoury et al. 2007), the transition from discovery to candidate health application involves identification and definition of the health condition, research on its biological and genetic associations and contexts, identification of potential interventions, and collection of data for Phase I and II clinical trials.

Each aspect of research at this phase is crucial, but at no point is it necessary to consider the perspectives of affected individuals and their communities and wider contexts (beyond genomic research considerations), let alone what we mean by recognition. Under the framework of responsive justice, by contrast, recognition will have a role to play throughout the translational cycle. At initial assessment and priority setting, underserved communities would be consulted about their own perceptions of health priorities, and about the research questions that are likely to aid in meeting those priorities. Recognition at this first step is crucial to the translation process; it is the context and grounding for research that can be not only efficacious but also just. At the discovery and development phases, ensuring recognition might require even more significant restructuring of the typical research system.

Community-based participatory research (CBPR) is one strategy that could be used to ensure the recognition element of justice in the discovery and development phases of research. Rather than waiting for researchers to approach communities with detailed, preplanned research agendas, as is common practice now, CBPR recommends that community groups be included in the earliest phases of research design and implementation. As such, community members are *collaborators* in research rather than mere *subjects* of research. This form of collaborative research is no doubt more difficult to carry out than the status quo, and it requires time to build relationships and to engage with community members, along with a willingness to learn and flexibility in focus. Researchers may discover that while they are interested in pursuing the genetic components of a major health concern in the community (obesity, for example), the community is worried about environmental contributions to the same condition. But more than simply shelving their research plan in favor of the community's proposed focus, researchers engaged in CBPR practices must be willing to let their core assumptions be challenged. Partnering with communities can affect both the quality of the data collected and the uptake of the intervention, because data collection strategies and dissemination means can be adapted to be more in line with community practices (AHRQ 2002; Baron et al. 2009). In one example, community advisors informed researchers that participant recruitment and the integrity of data could be compromised if there were not sufficient safety nets (e.g., alternative housing) for participants in a study about lead poisoning levels in low income housing (Jordan, Gust, and Scheman 2005).

Recognition at this level will require structural changes. One example is moving toward collaborative approaches to building positive relationships that facilitate research to encourage both short-term and long-term benefits to communities, particularly if research is focusing primarily on a time point further down the pathway. In some instances, knowing that anticipated benefits from translational research may be several years away, researchers can help to meet community needs through education, skill-based workshops, or creating infrastructures and capacity for research. It is possible that the individuals best placed to provide such services may not be genetics researchers per se but rather social scientists or others who have greater expertise in this area and who work in partnership with genetics researchers (e.g., nutritionists, community psychologists, educators). Through recognition, researchers can work to transform standard research practices into responsive practices by attending to both benefits and potential harms of the research from the community perspective.

Other structural changes might also need to take place. The relatively short grant timelines that are typical in academic research do not allow for the time required to build relationships with underserved communities, particularly where distrust of the research establishment is widespread and built on a history of abuse and/or neglect. Furthermore, what counts as a "deliverable" for a grant might need some redefinition, such that meetings in the service of relationship building can be counted as progress toward study aims. Budgets would need flexibility to support the foundational work for CBPR, as well as the possibility of supporting new avenues for research discovered through the CBPR process.

Attention to recognition and distribution is necessary to ensure justice along the translational pathway. Yet not every marginalized group will have sufficient organization to press its claims of misrecognition, and even when they do, solutions crafted in collaboration with researchers may require revision over time. This latter concern is particularly salient given that long-marginalized communities may be vulnerable to exploitation by more powerful interests; the offer of aid and collaboration may be quickly taken up even if the conditions of that collaboration are not fairly crafted. That is why, in addition to distribution and recognition, responsive justice also requires responsibility

Responsibility

It might sound odd to include responsibility as an *element* of justice, rather than a separate moral requirement that serves to ensure that justice is done once it has been identified. But in our model of responsive justice, responsibility plays a significant role in determining what justice is. *Responsibility highlights the role of the researcher in identifying his or her obligations to direct his or her work toward ends that have benefit beyond his or her own sphere.* For the researcher who might accept that recognition and distribution are important to consider all along the translational pathway generally, but who thinks such considerations are not particularly relevant to his or her work, we make the following case: because the traditional structure of research in which the individual participates, and which elevates his or her social status, material circumstances, and power, itself contributes to the marginalization of less privileged social groups, the researcher is materially linked to that marginalization to some degree. His or her actions, in other words, contribute to the injustice. Acknowledging a position of relative privilege, and the ways in which the system that rewards and supports that researcher simultaneously harms or neglects others, is part of taking responsibility. By attending to this responsibility, even where claims of misrecognition may not yet have arisen from marginalized groups, the researcher helps to identify potential injustices and shapes his or her work in response to them. Thus, when we insist on the importance of raising questions about responsibility at each stage of the pathway, we aim to help researchers rethink the scope of their obligations.

Descriptively, human beings find themselves, always, in relation to other persons—acting, reacting, revising, and responding. Because we are never free from the context of connection, many of us experience a sense of being "haunted" by others and their hold on us. We sense that people who are worse off than we are through no fault of their own—whether by accident of birth, history of discrimination and stigma, and so on—deserve a better lot in life. And when we are relatively privileged, particularly as academic scholars and researchers, we feel haunted by that fact because we know that we could always do more in the struggle for justice. Of course, we often rationalize explanations for our relative positions of privilege and/or allow our attention to be turned away from the needs of the vulnerable. Yet one can never escape the fact of connection, or the demands of responsibility for the other. This can mean that I, whoever I am—scientist, bioethicist, policy maker—will feel that something (someone) is pulling at me, tugging at my conscience, and reminding me that my part in the struggle for justice is not finished. The cancer vaccine researcher discussed earlier in this chapter is an example of such responsibility; she was committed to producing a vaccine that could help a less advantaged population, and she allowed her research design to be shaped by that commitment.

Prescriptively, we argue that human beings have a responsibility to one another, and that those of us in positions of privilege or power have a heightened responsibility to and for those with less, especially when the benefits we receive from our positions of privilege are partly built on the backs of marginalized others. Obligations to benefit others are often seen as less stringent than obligations not to harm others, but sometimes privilege is garnered in a system that itself creates harms to others, even if those harms often go unnoticed due to practice norms. Thus, while basic science researchers scrambling for funding might not think of themselves as having any particular position of privilege, members of marginalized communities often see it differently.

The relative privilege of researchers *increases* their responsibility to such populations because of their position within an institutional system that has an attitude of neglect toward the vulnerable/medically underserved. While individual researchers may not feel they are contributing intentionally to the neglect of vulnerable populations, the very invisibility of their privilege tends to perpetuate inequities. On this view, failing to admit our privilege or acknowledge the ways in which dominant social structures have benefited us can lead to significant harms for others, not least of which is the erosion of trust. In their work looking at the responsibility of researchers to be trustworthy, Jordan et al. note the following:

> It is not rational to trust those who have a track record of disrespectfully treating members of a community you identify with . . . or who take no interest in what members of your community have to say to them or in the effects that their views about your community have on the people in it. Given the depth and pervasiveness of social, political, and economic inequality in

> the United States today, it needn't take malevolence or malfeasance for researchers to act in ways that give rise to such perceptions. Ordinary, orthodox scientific method is frequently sufficient. (Jordan, Gust, and Scheman 2005, p. 51)

In other words, individual researchers need not intend any neglect of marginalized populations; their active participation in a practice structured so as to maintain that marginalization and/or to treat it as an object of study does implicate them in responsibility.

If researchers can be assigned some responsibility for the marginalization of the underserved, what do they owe particular others *qua* their position as researchers, committed to improvements in human health and welfare? Howard Brody has written that physicians and health care providers have a responsibility to acknowledge the power they have, to embrace the goal of sharing it with patients, and to direct their efforts toward morally justified ends. In his words: "We can have the highest degree of confidence that the healer's power is being used ethically and responsibly when that power can be described as owned power, shared power, and aimed power" (Brody 1993, p. 43). Such an approach makes sense for researchers as well as clinicians; their power should be acknowledged, shared with marginalized groups to ensure that the research aims are informed by their concerns, and aimed at ends beneficial for the least privileged.

Questions of responsibility are most obvious at the assessment and priority setting step, when an investigator or funder is choosing to focus on a particular issue or population. For instance, we can envision some researchers focusing primarily on potential individual benefits that could result following a successful clinical trial: improved health for those who can afford the drug or therapy, academic and business success for those who shepherded the new product through the approval process, and financial gain for those who hold market shares in it. For such researchers, considerations beyond the delivery stage might appear irrelevant. Yet the element of responsibility highlights the salience of outcomes research: when such "successful" interventions are integrated into practice, do they in fact provide gains in public health, particularly for populations most affected by the condition to be treated? Responsible researchers will foreground human health impact at any phase in the translational cycle, and they will ensure that their work is designed and handed off in a responsible manner, because of the critical reflection required by the assessment and priority setting step. The full job of translation requires assurance that health is indeed improved, rather than simply making a wider range of options available to the relatively affluent groups who are already relatively healthy.

As one example, responsible researchers might want to advocate for changes to the existing incentive structure for pharmaceutical companies. What if pharmaceutical companies could be offered rewards on the basis of something other than what the market will sustain, particularly given that many individuals are not able to afford payment for medicine? Philosopher Thomas Pogge and economist Aidan Hollis have proposed a Health Impact Fund that would be supported by national governments on the basis of a small percentage of gross national income and distributed to pharmaceutical companies on the basis of positive impact on global health (IGH 2009) Such a fund would offer financial incentives to attract companies' attention away from profitable "me-too" lifestyle drugs for hypertension or erectile dysfunction to products such as vaccines, or to more effective treatments for malaria, diarrhea, or other common but deadly health concerns in the global south. This kind of proposal might begin to address concerns raised in Chapter 5 by Patricia Deverka and David Veenstra regarding the difficulty of balancing the cost of innovation with the need for solid evidence of clinical utility in the development phase. For us, responsibility to the other, particularly where one's actions contribute directly to the harm or neglect of the other, is an element of justice that cannot be ignored.

GENETICS AND JUSTICE: TRANSFORMING RESEARCH PRACTICES

This book takes up the task of illustrating research practices at every stage of the translational pathway: assessment and priority setting, Discovery, Development, Delivery, and Outcomes. The chapters that follow are replete with case examples that show the work, focus, value judgments, and pressures that occur at each phase of research. Further, we offer commentaries after each section to highlight how standard research practices can be transformed by taking response justice considerations seriously.

In Chapter 2, legal scholar Patricia Kuszler lays the groundwork for better understanding the social, political, and economic forces that currently drive the process of genetic innovation and translation. Kuszler explains how alterations to funding streams, technology transfer and intellectual property law, and market forces have created a "formidable status quo" for standard research practice. This entrenched infrastructure poses numerous barriers to the potential for medical advances due to genetic innovation to benefit currently underserved populations. Kuszler explores the context of translational research, to better understand how the current situation came to be, and what forces must be reckoned with to alter the context in the name of justice.

The next two chapters center on discovery. In Chapter 3, population-geneticist-turned-bioethicist Stephanie M. Fullerton explores what she calls the "input/output" problem—namely, what happens when most of the genetic samples used in discovery research come from a particular population group (European race/ethnicity). Fullerton presents three case examples to show how individuals of non-European race/ethnicity may fail to benefit from established and widely marketed genetic tests.

The second discovery-phase chapter (Chapter 4) offers a case example of how advocates can bring about changes in basic science practices, with the aim of encouraging effective translation. Genetic epidemiologist Holly Tabor and medical sociologist Martine Lappé relate the story of the Autism Genetics Research

Exchange parent group, founded to foster research aimed at producing biomedical treatments for autism. Although no treatment has been forthcoming, the story demonstrates how advocates can influence the traditional scientific values (e.g., nonsharing of data among competitors) to encourage a more collaborative research network.

The next two chapters of the book focus on the development phase of research. In Chapter 5, health policy analyst Patricia Deverka and pharmacogeneticist David Veenstra show how difficult it can be to amass enough evidence to demonstrate clinical utility for promising pharmacogenetic innovations. They highlight some of the negative consequences that result from a combination of minimal regulation and interests in profit. Using the example of warfarin, an anticoagulant for which the appropriate dose can be determined through genetic testing related to metabolic uptake, Deverka and Veenstra uncover the reasons why a promising innovation may languish in development, and they consider possible remedies for this problem.

Chapter 6 features a review of the history of genetic testing in the reproductive realm by medical anthropologist Nancy Press, pediatric bioethicist Benjamin Wilfond, geneticist Mitzi Murray, and internist and geneticist Wylie Burke. The authors illustrate the ways in which provision of information has replaced the focus on clinical utility in defining benefit. They show how discussions about the value of prenatal testing and the reasons given for expanding its reach are often silent about abortion—a common consequence of positive diagnoses in prenatal testing. They express concern about how this understanding of information as sufficient benefit to justify testing is now expanding into the realm of personalized medicine, with potentially troubling effects.

Delivery of genetic medicine relies on effective placing of the innovations within clinical practice guidelines, and this is the focus of the next section of the book. In Chapter 7, medical geneticists Anne-Marie Laberge and Wylie Burke discuss the process of practice guideline development and point to challenges that arise when evidence is limited or lacking, as is often the case with genetic medicine.

The chapter by public health genetics researcher Nora Henrikson and Wylie Burke (Chapter 8) highlights the kinds of trade-offs that inevitably must be made in designing health care delivery systems. They argue that in the current fragmented American health care system, implementation of many genomic innovations will serve only to increase health disparities. Henrikson and Burke propose the concept of a health commons that would explicitly recognize limited health care resources and emphasize an expanded understanding of benefit. They urge more inclusive public input in discussions about the allocation of health care resources. In addition, they suggest that representation from medically underserved groups would likely lead to a greater emphasis on meeting existing standards of care over the implementation of expensive high-tech expensive genomic innovations.

Attention to outcomes research, the subject of the next section of the book, has been very limited. In Chapter 9, public health genetics researcher Catharine Riley and health economist Carolyn Watts tell the story of the development of newborn screening to show how advocacy groups pushed for widespread access to early screening but did not necessarily follow up to evaluate the relevant outcomes. The lack of advocacy for outcomes research raises concerns in relation to recent pushes for expansion of newborn screening panels, despite a lack of evidence about the usefulness of doing so.

Chapter 10, by Wylie Burke and Nancy Press, focuses on genetic testing for breast cancer and questions whether this now widely available testing has had the hoped-for health outcomes. They point to disparities in access to the testing as well as limited relevance of *BRCA* testing for most women who experience breast cancer. They take up the more general question of what a just genomic approach to health disparities might look like when it prioritizes the needs of medically underserved populations.

Chapter 11, by one-time cell biologist and current community-based researcher and educator Rosaline James and health services researcher Helene Starks, returns to assessment and priority setting. They tell the story of how the problem of youth suicide in Alaska Native communities gets framed in the early stages of research, and they raise concerns about who does the framing. Some researchers approach the problem from the perspective of genetics, hoping to identify genes related to depression or other suicide-linked conditions; others focus on cultural and behavioral interventions that target social determinants of suicide. As James and Starks show, priority setting should involve collaboration between all the relevant stakeholders—particularly including the target communities—if the interventions are to be effective.

Interspersed between the sections on the phases of research are commentaries that aim to highlight the central elements of responsive justice—redistribution, recognition, and responsibility—that arise in each phase. By proceeding this way, following genetic research practices through a translational cycle aimed at public health benefit, our goal is to demonstrate how our recommendations are grounded in actual innovations and practices rather than theoretical abstractions. Our hope for the book is that genetic researchers might find a view of the possible for themselves, that funders might see opportunities, and that members of the general public can recognize and demand research that is carried out responsively and responsibly. Though our focus is on genetics, primarily in the United States, we believe this approach can be extended to other kinds of research and applied in other contexts.

REFERENCES

[AHRQ] Agency for Healthcare Research and Quality. (2002). Community-Based Participatory Research: Conference Summary. Agency for Healthcare Research and Quality Web site. http://www.ahrq.gov/research/cbpr/cbpr1.htm Updated July 2002. Accessed November 2, 2010.

Baron S, Sinclair R, Payne-Sturges D, et al. (2009). Partnerships for environmental and occupational justice: contributions to research, capacity and public health. *Am J Public Health*. 99(Suppl 3):S517–S525.

Brody H. (1993). *The Healer's Power*. New Haven, CT: Yale University Press.

Burke W, Kuszler P, Starks H, Holland S, Press N. (2008). Translational genomics: seeking a shared vision of benefit. *Am J Bioethics*. 8(3):54–56.

Contopoulos-Ioannidis JP, Ntzani E, Ioannidis JP. (2003). Translation of highly promising basic science research into clinical applications. *Am J Med*. 114:477–484.

Dougherty D, Conway PH. (2008) The "3T's" road map to transform US health care: The "how" of high-quality care. *JAMA*. ;299(19):2319–321.

Fraser N, Honneth A. (2003). *Redistribution or Recognition: A Political-Philosophical Exchange*. London: Verso.

Gamble V. (2006). Trust, medical care, and racial and ethnic minorities. In: Satcher D, Pamies R, eds. *Multicultural Medicine and Health Disparities*. New York, NY: McGraw Hill; pp. 437–448.

Goering S, Holland S, Fryer-Edwards K. (2008). Transforming genetic practices with marginalized communities. *Hastings Cent Rep*. 38(2):43–53.

[IGH] Incentives for Global Health. (2009). Incentives for Global Health Web site. Yale University Web site. http://www.yale.edu/macmillan/igh/. Updated November 2009. Accessed November 23, 2009.

Jordan C, Gust S, Scheman N. (2005). The trustworthiness of research: a paradigm of community-based research. *Journal of Metropolitan Studies*. 16(1):39–57.

Khoury M, Gwinn M,Yoon P, Dowling N, Moore C, Bradley L. (2007). The continuum of translation research in genomic medicine: how can we accelerate the appropriate integration of human genome discoveries into health care and disease prevention? *Genet Med*. 9(10):665–674.

[NIH] National Institutes of Health. (2001). NIH Policy and Guidelines on The Inclusion of Women and Minorities as Subjects in Clinical Research – Amended, October, 2001. http://grants.nih.gov/grants/funding/women_min/guidelines_amended_10_2001.htm. Updated October 1, 2001. Accessed June 21, 2010.

Reverby S. (2009). *Examining Tuskegee: The Infamous Syphilis Study and Its Legacy*. Chapel Hill, NC: University of North Carolina Press.

Rubio DM, Schoenbaum EE, Lee LS, et al. (2010). Defining translational research: Implications for training. *Acad Med*. 85(3):470–475.

Smedley BD, Stith AY, Nelson AR (eds.); Committee on Understanding and Eliminating Racial and Ethnic Disparities in Health Care. (2003). *Unequal Treatment: Confronting Racial and Ethnic Disparities in Health Care*. Washington, DC: National Academies Press.

Starks HS, Edwards K, Kelley M, Burke W. (Forthcoming). Expanding the concept of the translational research.

Sung NS, Crowley WF, Jr., Genel M, et al. (2003). Central challenges facing the national clinical research enterprise. *JAMA*. 289(10):1278–1287.

Westfall JM, Mold J, Faqnan L. (2007). Practice-based research—blue highways on the NIH road map. *JAMA*. 297(4):403–406.

Woolf S. (2008). The meaning of translation and why it matters. *JAMA*. 299(2): 211–213.

2

The Social, Political, and Economic Underpinnings of Biomedical Research and Development: A Formidable Status Quo

PATRICIA C. KUSZLER

As we consider the goal of translating advances in genetic research into improved individual and population health outcomes, questions arise about how this goal can be achieved within the current social, political, and economic research infrastructure. This chapter explores the underpinnings of biomedical research and development in the United States and the powerful forces that have shaped and continue to shape the national research agenda. Most of these forces disfavor historically underserved populations and will, indeed, present a formidable barrier to their realization of benefits from genetic science.

This chapter begins with a discussion of the funding streams that support the research enterprise in genetics and other biomedical sciences, as well as the social, economic, and political incentives that fuel funding decisions. Second, it considers the impact of technology transfer laws and intellectual property rights on the research and development of drugs and other medical products. These laws have fundamentally altered the research agenda during the past quarter-century. Third, it explains how the rigorous regulatory review process contributes to the high costs of drug and biotechnology development. Market imperatives, patent protection laws and technology transfer incentives, and the high cost of regulatory review all work against a research agenda that addresses the needs of disadvantaged populations. These challenges already have arisen with respect to research in and development of genetic technologies.

RESEARCH FUNDING: WHERE DOES THE MONEY COME FROM?

Funding for genetics research, as for the entire universe of biomedical research, can be drawn from several potential sources, including federal and state governments, private pharmaceutical and biotechnology firms, private foundations, and donors. As health research moves from the lab into clinical practice, health insurers and health plans, both public and private, become funders of clinical deployment.

For much of the twentieth century, the primary funder for biomedical research in the United States was the federal government. Prior to World War II, government funding for research was a modest half-million dollars a year; in the postwar era, the federal government invested heavily in biomedical research through the National Institutes of Health (NIH), ushering in what is now termed the "golden age" of research. That growth has continued in more recent years: NIH funding reached $14 billion in 1997 and had doubled to $28 billion by 2004 (NIH 2010). Through its intramural and extramural funding programs, NIH poured resources into virtually every aspect of biomedical research during the last half of the twentieth century. The Human Genome Project received some of this largesse with the hope that unlocking the genome would revolutionize understanding of human disease.

The Human Genome Project began in 1988 as a congressionally funded collaboration between the U.S. Department of Energy and the NIH. Its scientific goals were to identify the genes present in human DNA, determine the sequence of its component base pairs, develop resources for storing and analyzing the data, and facilitate the transfer of this new knowledge and technology into the private sector for product development. In 1989, the National Center for Human Genome Research was established to carry out the NIH role in the project. The next year, the Human Genome Project was officially launched. Shortly thereafter, the Wellcome Trust, a global nonprofit funder of health research based in the United Kingdom, became a collaborator, and over the next several years many other nations and organizations joined the growing international consortium. The Human Genome Project was originally planned to last for 15 years, but early successes led to an accelerated timetable. By 2003, two years ahead of schedule, the human genome had been completely sequenced.

In 1997, the National Center for Human Genome Research was renamed the National Human Genome Research Institute (NHGRI) and elevated to full institute status within the NIH, joining 26 other permanent institutes and centers (NHGRI 2010). Federal funding for the human genome project grew from $27.9 million in 1988 to $484 million in 2008 (U.S. Department of Energy Office of

Science 2004). The history of the National Human Genome Project and NHGRI illustrates some of the factors that define worthiness for federal funding.

Through NIH and other federal grant-making agencies, the U.S. government invests public resources in health research with the ultimate goal of improving health outcomes. Federal research funding generally reflects priorities identified by Congress, the Executive Branch, and the public and private stakeholders that mold public policy. Human genetics—with its exciting promise of unlocking the secrets of our DNA—quickly captured the interest of Congress as well as the international community. One could argue that the quest for the genome has some parallels with the space race, in that scientists around the world actively competed—and continue to compete—to be the first to unravel the secrets of the human genome. This competitive model, in which a specific scientific quest is identified and highlighted for intensive attention, has a rich political history replete with several illustrative examples.

One excellent example is the much-heralded War on Cancer declared by President Nixon and signed into law with the National Cancer Act of 1971. At a time when the nation was divided and dismayed by the unpopular war in Vietnam, it was united in its common desire to conquer cancer, a disease that was then viewed as a virtual death sentence. Over the next 30 years, more than $200 billion was dedicated to cancer research, with much of those funds devoted to research on the biologic processes of cancer. Although 1.5 million scientific papers on cancer have been published to date, progress in treating and curing cancer has been extremely variable, with a few stellar successes—such as childhood leukemia—but an overall poor profile when compared to improved outcomes in other research areas (e.g., treatment of cardiovascular disease). Nevertheless, ardor for the War on Cancer is undiminished. The fact that successes have been relatively few has not deterred funding or quelled interest in cancer research. Nearly every U.S. president since Nixon has publicly proclaimed and reaffirmed the goal of curing cancer (Begley 2008).

The public can play a role in setting the research agenda as well, as demonstrated by a more contemporary example. As the AIDS epidemic unfolded in the early 1980s, the first group affected was the gay community. Dissatisfied with the federal government's response to this devastating public health problem, the gay community founded the AIDS Coalition to Unleash Power (ACT UP). ACT UP waged a confrontational, sustained, and successful battle to intensify research efforts, increase funding, and decrease the regulatory timeframe to get drugs and interventions to patients afflicted with HIV/AIDS (Hoffman 2003). Through this aggressive advocacy, a stigmatized but relatively wealthy, educated, and politically well-connected group captured the attention of the scientific establishment and successfully refocused the agenda for government-funded research.

However, the major force in research funding for the twenty-first century is the private sector. Over the past two decades, government funding was first supplemented and then surpassed by funding provided by private industry and international pharmaceutical and biotechnology firms. Following the dramatic growth of government funding in the late twentieth century, federal appropriations for research markedly slowed after 2004. Indeed, in recent years, federal funding has grown only about 3% per year, an increase barely matching inflation (Mandell 2004).

In contrast, the proportion of research that is paid for by pharmaceutical companies, medical device manufacturers, and other private-sector industry has increased at a remarkable rate over the same period. In 1970, the private sector accounted for $10.4 billion in research funding—roughly 20% of total research funding. In 1997, the private sector spent $133.3 billion on research, more than twice the amount spent by the federal government (Rampton and Stauber 2002). In sum, although government funding of research has incrementally increased over time, private funding has increased exponentially. The private sector is now the dominant funder of biomedical research, and this sector has changed markedly over the past two decades.

The private sector is characterized by ever-larger pharmaceutical conglomerates. Firms have been consolidated through mergers and acquisitions, resulting in larger corporations that are increasingly divorced from the scientists and public health needs—and a transition away from "pure" science to science that has rapid application to product development. These large corporations are driven primarily by the quest for profit and the need to both perpetuate their enterprises and guarantee their shareholders a return on their investment. The result is a "blockbuster mania" that demands research that can generate reliable and dramatically large profits. The goal is to develop products that will produce sales in excess of a billion dollars a year and, moreover, will provide those sales in a "front-end upswing" to maximize overall profit before any patent protection expires (Cuatrecases 2006).

This dynamic fosters the development of products that have a known, reliable, and large customer base. Such drugs and innovations typically mimic other successful products, leading to the penchant for "me too" drugs that provide a minor, but incremental, improvement to an already successful blockbuster drug. Successful entrants in the market usually are drugs and innovations that provide a lifestyle benefit for many, rather than a life-saving opportunity for a few. For example, drugs addressing arthritis and joint pain, erectile dysfunction, depression, and anxiety are enthusiastically developed, with the knowledge that they will be readily marketable in the United States, Europe, and other industrialized markets.

In addition to increases in government and private-firm funding, the past two decades have witnessed a growing role for nonprofit private foundations and public–private partnerships. The Bill and Melinda Gates Foundation is a leading example from the former category. Over the past several years, the Gates Foundation has poured well over $500 million—only slightly less than NIH over the same period—into malaria research, prevention, and treatment (Malaria R&D Alliance, 2005; Gates Foundation, 2006). An example of the public–private partnership model is the Global Fund to Fight AIDS, Tuberculosis and Malaria. Established in 2002, the Global Fund is a collaboration of private-sector corporations, nongovernmental organizations, and governments. It attracts, manages, and disburses resources to fight AIDS, TB, and malaria and has become the primary source of funding for such research (Global Fund 2010).

In addition to the large foundations and public–private partnerships, patient- and disease-focused advocates may play a role in molding the research agenda and influencing funding for research. Such advocates frequently address ethical and policy issues flowing from research initiatives and are increasingly included on NIH scientific review panels (Foster, Mulvihill, and Sharp 2006). Similarly, such advocates may also be represented in private foundation deliberations on funding. Involvement of such advocates in making funding decisions offers an opportunity, albeit a somewhat limited and peripatetic one, for such advocates to influence and guide research priorities. Notably, however, such advocates are not likely to be drawn from underserved populations that typically suffer from inadequate advocacy at every level.

Beyond influencing funding, there is also nascent involvement of patients' rights groups in shaping research by providing dedicated directed funding. For example, families affected by Canavan disease were critical in generating research to identify the mutation that causes the disease. Canavan disease is a gene-linked disorder that causes degeneration of nerves in the brain and usually results in death by age 4. Families provided not only DNA samples but also financial backing to researchers at Miami Children's Hospital Research Institute to identify the cause, with the ultimate goal of developing reliable prenatal and carrier testing (*Greenberg v. Miami Children's Hospital Research Institute, Inc.*, 264 F.Supp.2d 1064 [S.D. Fl. 2003]).

Despite their contributions, the families' ability to control the dissemination of the research and secure benefits for their group proved problematic. When the Canavan disease mutation was discovered, researchers moved quickly to patent their discovery with the aim of developing an economically lucrative commercial testing product. The Canavan families, whose aim had been to provide benefit to afflicted families, argued that both testing and further research would be burdened by licensing fees charged by the researchers. The families sued, arguing that the researchers' actions violated tenets of informed consent; constituted fraudulent concealment and a breach of fiduciary duty; amounted to conversion of property, given that the families' DNA was the critical substrate for the research; misappropriated a trade secret; and resulted in an unjust enrichment for the researchers. The court ultimately found that the Canavan families' rights had been abridged by the researchers, but not because their DNA had been appropriated; rather, the court found that the Canavan families' financial support of the research justified their claim that the researchers were extracting an unjust enrichment. The court clearly found the monetary investment to be a credible one, but did not view the genetic contribution similarly. This precedent would likely disadvantage those who can contribute genetic substrate to a research enterprise, but not financial backing.

Despite its origin as a government-funded initiative, genomics research in the United States reflects the current dynamics of research funding. Like other biomedical research, genomic research initially drew its funding from government sources but, over time, has become increasingly dependent on private funding sources.

The most comprehensive study of funding for genomics research was conducted in 2000 by the Stanford-in-Washington Program, whose World Survey of Funding for Genomics Research found that the private sector exceeded the public sector in funding genetics research. Although the exact quantification was impossible to derive, it was estimated that private-sector firms, most of which were based in the United States, spent at least $1 billion on genetics research (and more likely twice that), as compared with $845 million spent by government and nonprofit organizations. The survey concluded that the initial fruits of genomics research, be they therapeutic or diagnostic applications, would primarily benefit populations in rich countries (Cook-Deegan, Chan, and Johnson 2000).

In addition to the raw funding, technology transfer provisions and intellectual property laws also foster development of biomedical products, genomic and otherwise, that have commercial value and a prospect of profitability.

TECHNOLOGY TRANSFER LAWS AND PATENTS: DRIVING DEVELOPMENT AND DIRECTION

Despite the robust federal funding that energized biomedical research during the latter half of the twentieth century, Congress became increasingly concerned that not enough public benefit was being derived from the massive investment. Scientific discoveries often did not result in drugs and products that offered health benefits or other useful applications. This problem was partially an unintended consequence of the massive federal funding that characterized research during most of the twentieth century. During the first several decades of federal funding, the federal government took the position that research funded by federal dollars was, in effect, public property with the patent title firmly held by the government.

Development of products from such research required the developer to obtain a license from the government—an arduous and often ambiguous undertaking. The fact of government ownership diminished incentives for commercial exploitation of research and further development of products. Developers in the private sector reasoned that even if they chose to license the technology from the government, they would be unable to restrict competitors from using the same substrate. As a result, much of the scientific research and development done under the aegis of federal funding languished unused and undeveloped. Prior to 1980, less than 5% of the 28,000 patents held by the government had been licensed to developers (Arno and Davis 2001). Most federally funded research remained at the "bench science" stage and did not move from the laboratory into the development of useful biomedical products and therapies.

Government officials and scientists argued that innovation was trapped in a maze of bureaucracy and unclear title that chilled further development. Despite historic investment of public resources, biomedical research was failing to deliver either public benefit or a revenue stream that could be reinvested in the research enterprise (Arno and Davis 2001).

Congress sought to break down these barriers by enacting legislation that would leverage federal investments in research while simultaneously providing incentives to researchers and their universities. It accomplished this by passing a series of laws designed to facilitate the transition from basic science research to commercial products by giving researchers and research universities the opportunity to share in the ownership and profits of commercial products derived from their work. Moreover, these laws paved the way to a new model of research funding.

Although discussion of a development-friendly technology transfer policy began during the Kennedy Administration, plans did not come to fruition until the 1980s. In 1980, Congress passed the Stevenson–Wydler Technology Act (15 U.S.C. §§ 3701-3717), which sought to improve utilization of technologies created as a result of federal funding. This act essentially signaled the switch to a cooperative model of research by requiring federal laboratories to improve technology transferring capacity by enacting policies and procedures that would encourage development of scientific inventions. As result of this act, all federal laboratories have established Offices of Research and Technology Transfer that seek to license their scientific discoveries to developers.

The Stevenson–Wydler Act was paired with the Bayh–Dole Act (35 U.S.C. §§ 200-212), also passed in 1980, which opened up patenting and licensing opportunities for research universities and researchers. The Bayh–Dole Act allowed universities and other nonprofit organizations to retain title to inventions and products discovered or supported by federal funding and grants. Private and nongovernment laboratories quickly established technology transfer offices and expertise in order to avail themselves of these new opportunities. The Bayh–Dole Act also allowed federal agencies to grant licenses to private organizations to allow development of products from the research conducted in the federal laboratories of NIH and the National Science Foundation. This licensing capacity was enhanced by allowing the federal government to grant exclusive licenses that confer significant marketing advantages. However, the federal government retained what are termed "march-in rights," meaning that under certain circumstances (i.e., if warranted by concern for public health and safety, or if the licensee failed to use and disseminate the invention as agreed), the government could access exclusively licensed technology.

Augmenting the Bayh–Dole and Stevenson–Wydler Acts is the Federal Technology Transfer Act of 1986 (15 U.S.C.A. §§ 3701-3714). This act allows federal agencies to enter into joint venture Cooperative Research and Development Agreements (CRADAs) with private research facilities. CRADAs provide an opportunity for the government to partner with private industry by contributing personnel, facilities, and equipment to a research endeavor while the private entity provides funding and other resources. The end result of these joint ventures is that the private entity has first access to licensing opportunities. Once again, the federal government retains the right to use the product and subvert an exclusive license under a march-in provision.

The revenue realized by universities from the shift to a public–private cooperative model has been substantial. In 1980, royalty payments to universities totaled a modest $1 million; by 1994, universities were netting more than $265 million in royalty payments (Blumberg 1996). Research universities are increasingly party to entrepreneurial agreements intended to provide them with income and new collaborative research and development opportunities (Wadman 2008). That said, while there are a few storied money-maker innovations, the majority of universities enjoy only modest returns from technology transfer efforts.

Universities have also sought to gain from research discoveries and technologies by more aggressively patenting and marketing the inventions themselves. Following passage of the Bayh–Dole Act, the number of academic institutions seeking and receiving patents increased rapidly. Not only were more universities receiving patents; the number of patents per university also increased substantially, from 589 in 1985 to 3,259 in 2003 (National Science Board 2006). There is a marked concentration of university-held patents in the life sciences, especially in biomedical areas; this concentration is not mirrored by U.S. patents in general.

Universities have exploited this increase in patent accumulation by investing in and expanding their technology transfer programs and negotiating an ever-increasing number of licensing and royalty agreements. In keeping with the life sciences/biomedical trend, pharmaceutical, biotechnology, and medical businesses are typical licensees of university-patented technologies. University income from patents and licensing reached $482 million in 1997; this was a six-fold increase from the 1989–1990 income of $82 million. The National Science Board has noted that all indicators show an accelerating use of patenting and technology transfer by universities.

Individual researchers also profit from aggressive patenting and the transfer of technology, and many take an active role in this process. The individual research scientist often triggers the cascade of entrepreneurial activity with respect to a new discovery or technology. One typical mechanism is a "faculty start-up company." In such arrangements, a member of the university research faculty is urged to recognize promising discoveries early and partner with the university to protect the intellectual property and its potential financial rewards. Both the scientist and the university will have equity in the start-up enterprise. This new partnership will seek to license the new technology to an industry player who will then develop the technology into a marketable product (Harrington, 2001).

The potential profits from such ventures result in researchers and universities seeking to focus their research efforts on applied research rather than basic bench research. There is a cogent argument that this new focus on the market and profit perverts the goals of "pure" scientific research (Korn 2000). Academic researchers are encouraged to patent and protect their inventions as soon as they recognize a discovery. This early patent protection forecloses research by others that might have led to an exponential increase in knowledge.

This refocusing of priorities has resulted in some researchers and universities being more closely aligned with private industry in terms of interests and incentives. Like private pharmaceutical and biotechnology firms, researchers and universities are drawn by the potential profits that their research product could generate. Researchers' career trajectories may be increasingly influenced by their ability to produce science that can be commercialized by an industry partner, to their own benefit and that of their academic institution. This is reflected in the proliferation of scientific articles that are coauthored by academic researchers and industry scientists. Between 1981 and 1995, such coauthored articles increased from 21.6% to 40.95% of published scientific articles (Rampton and Stauber 2002).

The technology transfer statutes have been a major boon to some researchers and universities as well as to private pharmaceutical and biotechnology firms. Entrepreneurial private development has flourished, and a small number of universities have gained impressive income (Wadman 2008). This phenomenon has been particularly marked in genetics and biotechnology. There has been vigorous patenting of DNA-related scientific discoveries before and during the Human Genome Project. A survey done in 2000 found that the U.S. government owned nearly 500 patents, private industry owned more than 2,700, and universities likely held even more. Although the catalog of DNA-related patents is difficult to fully ascertain, it is clear that ownership of DNA-related patents is heavily concentrated in the United States (Cook-Deegan, Chan, and Johnson 2000).

The move to patent as early as possible in the course of research is accompanied by another trend: to patent as broadly as possible, so as to secure rights to as much of the scientific territory as one can. Researchers and developers reason that this aggressive approach secures for them a second generation of invention and products. In addition, it hinders others from easily building upon the earlier science to make new discoveries. If an aspiring scientist must pay a licensing fee to use an earlier scientific process, he or she will be less likely to pursue that particular vein of research.

There have been a few attempts by members of underserved groups to counter the incentives of the patent and technology transfer system. In the aforementioned case, the Canavan families were outraged when the researchers patented their discovery, thus setting themselves up to be reap the financial yield of the resulting test as well as licensing fees for subsequent use of the scientific process. Patient advocacy groups, particularly those dedicated to genetic maladies, have sought to enhance their role in patent ownership by conditioning use of their genetic materials on a share of the patent rights.

There is ongoing legal and human rights controversy about how technology transfer and patent laws can be structured to respect the rights of inventors and patent holders while allowing inventions to be made available to needy end users, be they disadvantaged groups within the United States or populations in the developing world. Patents covering most of the innovations in biotechnology and genetics are owned and controlled by corporations and governments in the developed world. The 2001 Human Development Report produced by the United Nations Development Programme noted that member countries of the Organisation for Economic Co-operation and Development (OECD) account for 19% of the global population but 91% of all new patents issued in 1998 (United Nations Development Programme 2001).

At present, patent protection and technology transfer imperatives present a significant barrier to the development of drugs and other innovations that would meet the needs of the underserved. The patent and technology transfer laws foster research that will generate a quick profit, usually through the development of drugs, devices, and biologics (such as vaccines) that can be marketed to large, relatively wealthy patient populations. Biomedical and pharmaceutical industry firms argue that seeking patent protection is essential to recoup the high costs of getting products through the regulatory approval processes.

THE HIGH COSTS OF REGULATORY REVIEW

The regulatory approval regime for drugs, biologics, and medical devices is costly and time consuming. The regulatory morass adds to the economic and marketing factors that make the biomedical industry reluctant to invest in the development of any product with uncertain profitability.

Biomedical firms face difficult and complex regulatory environments as they seek government approval to market their products. This is particularly true in the United States, but other industrialized countries also require compliance with onerous approval processes. Moreover, until very recently, there has been little or no reciprocity among nations with respect to approval processes for drugs and vaccines. Recent efforts to harmonize the regulatory processes have resulted in a proposed regime that mimics the regimes in the United States, the United Kingdom, and Japan. Critics argue that the proposed process requires the bureaucratic and scientific underpinnings of the high-income, industrialized nations.

The U.S. Food and Drug Administration (FDA), nicknamed the "foot-dragging and alibi" agency by its many detractors, is viewed as having the most conservative, arduous, and time-consuming of the world's drug approval processes. It has been accused of "deadly over-caution" in the lay press; this is not a wholly undeserved label.

First of all, the FDA has more than one set of rules. It approves and regulates drugs using one process mandated by the Food, Drug, and Cosmetic (FD&C) Act (21 U.S.C. §§ 301-393). Biologics such as vaccines fall under a regulatory regime governed by the Public Health Service (PHS) Act (42 U.S.C. § 262). Finally, medical devices are regulated under the Medical Device Amendments (MDA) (21 U.S.C. § 360) and its subsequent amendments. Adding to the complexity of three separate regulatory regimes are ambiguities with respect to product definition. For example, some biologic products may be deemed to have drug-like behavior and be susceptible to regulatory approval under both the drug and biologic regimes. Some products, such as laboratory-developed genetic tests, may be subject to no regulation; alternatively, if the testing is provided as part of a test kit, it may be deemed a device and subject to regulation as such.

The complications presented by the current system can be readily illustrated by the least complicated of the regulatory regimens: the approval process for a new drug. The FDA has rigorous safety and effectiveness standards for new drugs, which are arguably guaranteed by a lengthy four-stage approval process. The approval regimen proceeds from an initial preclinical testing phase performed on animal subjects, followed by an "investigational new drug" phase that requires three separate stages of clinical trials with human subjects. The accumulated data are then submitted to the FDA in a new-drug application for evaluation, review, and additional safety and effectiveness testing. This extended process routinely takes a decade or more to complete and is extremely costly for the pharmaceutical manufacturer. After approval has been granted, the fourth phase commences; it consists of post-marketing surveillance and monitoring of the drug's safety and efficacy. This fourth phase is largely dependent on self-regulation by the pharmaceutical firms that will be the recipient of adverse-event reports generated by health providers and patients. Phase four is the least well-developed part of the regime and is often characterized by indirect regulation borne through liability suits brought by those injured by the product.

Biomedical and pharmaceutical firms argue that FDA's requirements discourage new drug development. Unless a drug is projected to be a clear winner in the market, developers are reluctant to incur the research and development costs, plus the expense of the lengthy approval process. Contemporary estimates of the cost range between $543 and $803 million [in 2000 dollars] (DiMasi, Hansen, and Grabowski 2003).

In addition to much-delayed return on the biomedical/pharmaceutical firm's investment, the testing and approval process also regularly consumes half or more of the period of patent protection. Under patent protection, the manufacturer is able to protect a new drug from competitive copycat drugs for a relatively short time; when some of that time is lost to the approval process, manufacturers have even less time in which to recoup the costs of research and development. As a result, manufacturers choose to develop only drugs that yield a higher profit margin, and which do so within the first few years after their market introduction, while they are still safe from competition. Thus, the chances that drugs targeted at small or poor patient populations will be serious contenders for development are significantly reduced (Hogan 1995).

As early as the 1980s, it was evident that new drugs for rare diseases or diseases primarily affecting poor, uninsured populations were not being developed. Such conditions were tagged "orphan diseases"; these under-targeted conditions were considered abandoned because there was little interest or impetus to develop pharmaceutical treatments and remedies for them. Developing a drug for an orphan disease would not produce a profit for the increasingly profit-driven pharmaceutical manufacturer.

Congress acted to remedy this situation by passing the Orphan Drug Act of 1983 (21 U.S.C. § 360), through which pharmaceutical firms were provided with three separate incentives to develop drugs for under-targeted conditions. First, Congress provided for a substantial tax break: the pharmaceutical company could receive a credit against taxes (not simply a deduction) for half of the cost of the clinical trial during the time the designated orphan drug was seeking approval. Moreover, the credit could be claimed even if the product failed to make it to market. Second, the Act provided for additional government planning assistance to the company throughout the course of the clinical trials and approval process. Finally and most importantly, the Orphan Drug Act granted the pharmaceutical company seven years of market exclusivity, regardless of whether or not the new orphan drug was patentable.

The benefits of the Orphan Drug Act were subject to certain limitations. First, only in very select situations would more than a single drug be deemed an "orphan drug" for a given disease. In addition, the incentives, particularly the market exclusivity incentive, were made available only to a narrow class of prospective drugs: those developed for diseases affecting fewer than 200,000 persons in the United States. This narrow classification was somewhat broadened by a 1985 amendment to the act, which provided for orphan drug status if there were no reasonable expectation that the costs of development and manufacturing could be recouped from U.S. sales. Moreover, the market exclusivity provision would remain applicable even if the target population grew to exceed 200,000 people during the period of exclusivity.

Passage of the Orphan Drug Act did spur drug development. While only 10 drugs treating such "orphan diseases" had traversed the FDA regulatory maze to approval in the decade preceding the Orphan Drug Act, 87 new drugs were brought to market in the first decade after its implementation (Hogan 1995). However, the act has also been exploited by pharmaceutical firms, which have learned that they can divide a more common disease into discrete subcategories of disease or indications to qualify for the orphan drug designation. Another ploy is to seek orphan drug status for one indication (e.g., to treat a rare hormone disorder), even while realizing that the drug will have great utility in treating a much more common malady (e.g., infertility). Upon reaching the market, physicians and patients are made aware of the orphan drug's versatility, and physicians prescribe the drug off-label for the more common condition, which effectively allows manufacturers to "double dip." Many firms have encouraged off-label usage by sponsoring expert speakers at medical conferences and using other marketing techniques, with only a few attracting the attention of regulators seeking to curtail off-label use. Through these techniques, many pharmaceutical companies have managed to extract windfall profits, despite the small numbers of patients targeted, by setting high prices for these products. The Orphan Drug Act does not impose price controls or requirements for supply or distribution on manufacturers. Several legislative attempts to comprehensively address these problems have been unsuccessful.

DISADVANTAGED POPULATIONS IN THE MODERN MARKETPLACE

The economic, social, and political forces outlined above are all well entrenched and not susceptible to change. It can be argued that intellectual property protections are vital to sustaining the momentum of research and that technology transfer is an essential step in ensuring that scientific innovations move from the research laboratory to the marketplace. Likewise, the regulatory road to approval for marketing is critical to assuring the safety and efficacy of drugs and other technologies before they are released to a naïve and vulnerable public. Attempts, such as the Orphan Drug Act, to encourage firms to address the needs of the disadvantaged and underserved have proved to be a two-edged sword, with the industry quickly learning how to game the system to enhance profits.

Given the dynamics of the marketplace and the incentives that are currently in place, dedicating limited resources to the development of products that can quickly earn a substantial return on investment is a rational and necessary choice on the part of researchers and industry. The net result of this unfortunate reality is that underserved populations, diseases afflicting small populations, and diseases that primarily affect people in the developing world receive little attention, save for the efforts of the Gates Foundation, the Global Fund, and other emerging models of private–public partnership. In part through their efforts, the concept of the "orphan disease" is increasingly being broadened to include not only undertargeted conditions but conditions common in the developing world that have not been focused on in traditional research and product development efforts. For example, the European Union's version of orphan drug legislation also applies to tropical and other diseases prevalent in the developing world.

It is likely that the effects of these barriers will be felt more keenly in high-technology research areas, such as genetics. Such high-technology areas typically require sophisticated and expensive research and clinical infrastructures, as well as a relatively sophisticated and wealthy end-user consumer population. This is already evident as we look at the genetic advances already available in the marketplace.

Genetic testing and technologies are perhaps most commonly seen in clinical practice today in the area of prenatal and newborn screening. Many of these tests and technologies have been in the marketplace for many years, and there is indeed a robust market. In the context of prenatal testing, a growing number of patients undergo an array of prenatal testing with the aim of early diagnosis of chromosomal and genetic disorders. These technologies are relatively inexpensive and easily available in most urban markets. However, access to these technologies requires access to prenatal care, which many women do not have due to lack of health care coverage, geographical distance, and socioeconomic factors. In industrialized nations, such underserved women will likely be poor and women of color. In developing nations, the majority of women will have no access to prenatal genetic testing opportunities.

More sophisticated testing, such as preconception/preimplantation genetic diagnosis, is only available to patients able to pay the substantial cost, as it typically is not covered by private, let alone public, health plans. Similarly, services such as genetic counseling may not be covered or may be covered only to a small degree. And both sophisticated genetic testing (pre- and post-implantation) and genetic counseling are typically found only in medium and large, relatively affluent urban areas where tertiary-care obstetric and medical services are in place.

Moreover, there is growing concern that the quest for a healthy child may transform into the quest for a genetically better child—or even one who, down the line and with more nuanced genetic testing, may be designed to have more socially desirable attributes. There is well-documented prenatal testing for sex selection in India and China, and recent research indicates that even in the United States sex selection is not uncommon (Almond and Edlund 2008). Much of genetic testing is unregulated, and there are growing numbers of firms offering an array of genetic tests that could easily be seized upon by consumers seeking to further hone their prenatal selection capacity. For example, susceptibility testing could be employed at the prenatal level to select against susceptibilities to cancers and other late-onset disorders. As the options in testing evolve, so too will the desire of prospective parents to ensure the best possible outcome. The result may be yet more stratification of society on the basis of genetic traits, yielding a society in which the underserved would likely be even further disenfranchised.

Testing outside the prenatal context evidences a similar pattern. For example, testing for breast cancer susceptibility is extremely expensive, largely due to the patent protection that Myriad Genetics, the test developer, currently enjoys. This testing has been aggressively marketed to physicians and to potential patients, but the price tag permits access only for relatively wealthy women. The aggressive marketing generates demand for a test that has limited clinical utility for most women (Mouchawar et al. 2005). As a result, women may struggle to pay for the test, when neither they nor their physician realizes that they are not appropriate candidates for testing. This is a perversion of the market in which the purchasers have incomplete or inadequate knowledge. Such perversions fuel increased health care costs, contribute to the growing unaffordability of health care coverage, and ultimately increase health disparities. In the case of breast cancer testing, these effects are intensified for minority women because of their reduced likelihood of receiving an informative test result, as discussed in Chapter 10.

Another example can be found in the promising area of pharmacogenomics, in which genetic testing is employed to choose and/or tailor the drug regimen chosen to treat a disorder. Pharmacogenomics research offers the possibility of providing more-targeted treatment for patients, by tailoring the medication or intervention to their specific genome. Already there is pharmacogenomic testing to determine whether a child is overly sensitive to a common chemotherapy drug and needs to have a lower dose. Scientists promise that in the future, genetic "designer drugs" will allow the patient maximal opportunity for a successful intervention. However, the benefit is predicated on the frequently false assumption that access to health care is available in the first place. If the child with leukemia has the misfortune to be living in Africa or in a rural area of the United States, the child may not even have access to initial diagnosis, much less nuanced pharmacogenomic testing to titrate treatment. The opportunity cost of developing such pharmacogenomic testing is high, and the ethical propriety of devoting massive resources to benefit a few, rather than devoting the same resources to benefit many, cannot be ignored. Unlike the orphan disease context, where we can incorporate neglected diseases into the paradigm, pharmacogenomics is geared toward the individual and is not easily translated to a population benefit. In the zero-sum game of limited global resources, the promise of pharmacogenomics and the ideal of personalized medicine may ultimately exacerbate disparities and worsen the lot of the underserved.

Although few genetic therapeutic interventions have been developed to date, there are some illustrative examples of successful treatments built upon genetics. One of these is the enzyme therapy Ceredase, which was developed as an orphan drug to treat Gaucher disease. This disorder results from an enzyme deficiency and leads to numerous liver, spleen, musculoskeletal, and central nervous system problems. In its most severe type, Type 1, the patient presents in childhood with severe effcts on these organ systems. As genetic knowledge accrued, scientists were able to identify the missing enzyme and eventually harvest it from placental tissue. In this example, we see some of the unintended consequences that tarnish the image of the orphan drug approach. A single year's supply of this enzyme, Ceredase, initially cost $350,000; a second-generation version, Cerezyme, which is produced using recombinant DNA technology, costs $200,000. Moreover, this enzyme therapy must be maintained over the lifetime. With ongoing therapy, individuals with this disorder can live relatively normal lives into and throughout adulthood. The astronomical costs generally are paid by the health plan—assuming that the patient is lucky enough to have private health care coverage or a generous Medicaid plan. Many families struggling to deal with Gaucher disease have been forced to "spend down" their assets to qualify for Medicaid (Kaiser Health News 2008). Faced with the negative public-relations impact of bankrupting many patents, Genzyme, the manufacturer of Ceredase and Cerezyme, ultimately initiated a program to provide the medication to those who could not afford it. Despite this, Cerezyme generated $840 million in sales in 2004, with a 90% profit margin for Genzyme (Medical News Today 2005).

In many ways, the plight of the underserved in the United States is mirrored on the global level. Virtually all of the same political, economic, and social factors that continue to disenfranchise underserved populations are at play with respect to research on neglected diseases plaguing the developing world. In fact, as previously mentioned, the European Union has recognized this parallel in its version of orphan drug legislation. Less than 1% of the 1,223 new medicines launched on the international market between 1975 and 1997 specifically targeted the class of tropical communicable diseases that forms the bulk of the neglected-disease category (Trouiller et al, 2001).

As we have seen, market forces inevitably skew research and development toward diseases that will yield high financial rewards; these choices are to the detriment of populations in developing countries (Schwab et al. 1999). This is one of the primary factors underlying what has been referred to as the global 10/90 disequilibrium: less than 10% of total global spending on health research is devoted to the diseases that account for over 90% of the global disease burden (Global Forum for Health Research 2009). While organizations such as the Gates Foundation have begun to address the 10/90 gap, the continuing inequity drives concerns that populations of developing and economically struggling countries are not receiving full and fair benefits of research and scientific discovery (Participants 2002).

In conclusion, underserved populations in the United States and around the world face a challenging future with respect to access to the fruits of scientific progress, including those derived from genetics research. The current economic, political, and social contexts dictate a research agenda that is directed toward the development of profitable products for the wealthy populations that can afford them. Moreover, the status quo is propagated by the complex web of expensive regulatory review and continued fortification of intellectual property rights.

Despite this, there are some approaches that speak to the needs of underserved populations and neglected diseases. Globalization is raising awareness of health disparities. This has led to renewed attention to the incentives borne in intellectual property protections and proposed broadening of public health exceptions contained in international treaties and agreements. Similarly, the EU approach to orphan diseases seeks to incentivize research that would address the needs of underserved populations, encompassing not only rare diseases but also neglected diseases. Finally, there is the growing role of private foundations and public–private partnerships in shaping the research agenda to promote and further global health. It remains to be seen whether these new approaches can counter, and conquer, the profit-making incentives that currently mold the biomedical research and development agenda.

REFERENCES

Almond D, Edlund L. (2008). Son-biased sex ratios in the 2000 United States Census. *Proc Natl Acad Sci U S A*. 105(15):5681–5682.

Arno PS, Davis MH. (2001).Why don't we enforce existing drug price controls? The unrecognized and unenforced reasonable pricing requirements imposed upon patents deriving in whole or in part from federally-funded research. *Tulane Law Rev*. 75:631–636.

Begley S. (2008). We fought cancer… and cancer won. *Newsweek*. September 6. http://www.newsweek.com/id/157548.

Blumberg P. (1996). From "publish or perish" to "profit or perish": revenues from university technology transfer and the § 501(C)(3) tax exemption. *Univ PA Law Rev*. 145(89):89–147.

Cook-Deegan R, Chan C, Johnson A. (2000). *World Survey of Funding for Genomics Research: Final Report to the Global Forum for Health Research and the World Health Organization*. http://www.stanford.edu/class/siw198q/websites/genomics/finalrpt.htm. Updated September 25, 2000. Accessed November 10, 2010.

Cuatrecases P. (2006). Drug discovery in jeopardy. *J Clin Invest*. 116(11):2837–2842.

DiMasi JA, Hansen RW, Grabowski HG. (2003). The price of innovation: new estimates of drug development costs. *J Health Econ*. 22(2):151–185.

Foster MW, Mulvihill JJ, Sharp RR., (2006). Investments in cancer genomics: who benefits and who decides? *Am J Public Health*. 96(11):1960–1964.

Gates Foundation. (2006). Global fund announces $500 million contribution from Bill and Melinda Gates Foundation. http://www.gatesfoundation.org/press-releases/Pages/aids-tb-malaria-global-fund-060809.aspxx 2006. Accessed December 29, 2010.

Global Forum for Health Research. (2009). "10/90 gap." http://www.globalforumhealth.org/About/10-90-gap. Updated 2009. Accessed November 10, 2010.

The Global Fund to Fight AIDS, Tuberculosis and Malaria. (2010). About the Global Fund. http://www.theglobalfund.org/en/about/?lang=en. Accessed November 9, 2010.

Greenberg v. Miami Children's Hospital Research Institute, Inc., 264 F.Supp.2d 1064 (S.D. Fl. 2003).

Harrington P. (2001). Faculty conflicts of interest in an age of academic entrepreneurialism: an analysis of the problem, the law and selected university policies. *J Coll Univ Law*. 27(4):775–831.

Hoffman B. (2003). Health care reform and social movements in the United States. *Am J Public Health*. 93(1):75–85.

Hogan JM. (1995). Revamping the Orphan Drug Act: potential impact on the world pharmaceutical market. *Law Policy Int Bus*. 26(2):523–561.

Kaiser Health News. (2008). High prescription drug prices lead to scrutiny of medication dosage levels. http://www.kaiserhealthnews.org/daily-reports/2008/march/17/dr00050990.aspx?referrer=search. Updated March 17, 2008. Accessed November 10, 2010.

Korn D. (2000). Conflicts of interest in biomedical research. *JAMA*. 284(17):2234–2237.

Malaria R&D Alliance. (2005). Malaria research and development: An assessment of global investment. http://www.malariavaccine.org/files/MalariaRD_Report_complete.pdf. Accessed December 29, 2010.

Mandel HG, Vesell ES. (2004). From progress to regression: biomedical research funding. *J Clin Invest*. 114(7):872–876.

Medical News TODAY. (2005). Wall Street *Journal* examines high price of Gaucher disease treatment Ceredase. http://www.medicalnewstoday.com/articles/33642.php. Updated November 17, 2005. Accessed November 10, 2010.

Mouchawar J, Hensley-Alford S, Laurion S, et al. (2005). Impact of direct-to-consumer advertising for hereditary breast cancer testing on genetic services at a managed care organization: a naturally-occurring experiment. *Genet Med*. 7(3):191–197.

National Science Board. (2006). Science and Engineering Indicators 2006. Appendix Table 5-68: U.S. Patenting Activity of U.S. Universities and Colleges: 1994–2003. http://nsf.gov/statistics/seind06/append/c5/at05-68.pdf. Updated February 23, 2006. Accessed November 9, 2010.

[NHGRI] National Human Genome Research Institute. (2010). http://www.genome.gov/27534788. Updated October 18, 2010. Accessed November 9, 2010.

[NIH] National Institutes of Health. (2010). NIH Almanac–Appropriations. http://www.nih.gov/about/almanac/appropriations/part2.htm. Updated May 14, 2010. Accessed November 9, 2010.

[Participants] Participants in the 2001 Conference on Ethical Aspects of Research in Developing Countries. (2002). Fair benefits for research in developing countries. *Science*. 298:2133-2134.

Rampton S, Stauber J. (2002). Research funding, conflicts of interest and the meta-methodology of public relations. *Public Health Rep*. 117(4):331–339.

Schwab K, Porter ME, Sachs JD, et al. (1999). *The Global Competitiveness Report 1999*. New York, NY: Oxford University Press, Inc.

Trouiller P, Torreele E, Olliaro P, et al. (2001). Drugs for neglected diseases: a failure of the market and a public health failure? *Trop Med Int Health*. 6(11):945–951.

United Nation Development Programme. (2001). Human Development Report 2001: Making New Technologies Work For Human Development. http://hdr.undp.org/en/reports/global/hdr2001/. Updated July 11, 2001. Accessed November 10, 2010.

U.S. Department of Energy Office of Science. (2004). Human Genome Project Budget. http://www.ornl.gov/sci/techresources/Human_Genome/project/budget.shtml. Updated September 14, 2004. Accessed November 9, 2010.

Wadman M. (2008). The winding road from ideas to income. *Nature*. 453(7197): 830–831.

3

The Input-Output Problem: Whose DNA Do We Study, and Why Does It Matter?

STEPHANIE MALIA FULLERTON

Discovery research—sometimes referred to as basic science—is an important first step in the development of health care innovations to benefit individuals and populations. In the genome sciences, discovery research typically involves the identification of genes and genetic variants that can be associated with specific health outcomes in reliable and replicable ways. Such known associations include increased susceptibility to certain diseases (e.g., certain familial cancer syndromes), one's likelihood of benefiting from a particular treatment (such as response to treatment for certain types of leukemia), or the risk of having an adverse reaction to medication (e.g., Stevens-Johnson syndrome with exposure to certain anticonvulsants).

Once such an association has been discovered, it might seem a straightforward matter to design a genetic test to identify individuals who may be at increased risk; however, developing such a test is harder than it looks. The simplest case is with certain types of genetic diseases studied at the level of individual families. For example, imagine that a woman with breast cancer decides to have genetic testing to determine whether she carries one of the known cancer-causing mutations in either the *BRCA1* or the *BRCA2* gene. If such a mutation is discovered, then testing for the presence of that specific mutation in other genetically related women in her family would be directly informative with respect to her risk of developing cancer. When discovery is based in population-scale investigation, however, most disease states or adverse drug reactions can be traced to mutations (either singly or in combination) in multiple genes. In other words, there are usually many different ways to perturb the developmental processes relevant to a given trait. Even for the class of so-called monogenic, or single-gene, disorders there typically are many hundreds, if not thousands, of mutations that can individually interfere with function of a given gene and lead to disease on a population scale. Test sensitivity (the ability to detect a predisposing gene variant when one is actually in play) is necessarily subject to the thoroughness with which these myriad genetic contributions are identified in early-stage discovery research. Thus, systematic biases in genomic discovery can lead to systematic biases in test performance, with important downstream consequences for the effectiveness of health care delivery and individual health.

This chapter focuses on the impact of one particular type of bias in genomic discovery research: bias with regard to population sampling. Varying rates of research participation across racial/ethnic groups, combined with background differences in genetic variation among populations, can negatively affect genomic translation. The ultimate effect of sampling biases may be that certain groups fail to benefit from the public investment in discovery science and interventions that may be based on those findings. In this chapter, three specific classes of discovery bias are examined, and potential policy remedies are discussed.

UNDERREPRESENTATION OF RACIAL/ETHNIC MINORITIES IN GENOMIC RESEARCH

One area for which there is consistent evidence of systematic bias in genomic discovery is the population distribution of study samples. The overwhelming majority of genetic effects have been characterized first, or only, in populations of European race/ethnicity. Although this bias is widely recognized, there have been only a handful of attempts to comprehensively summarize rates of participation of different racial/ethnic groups in genetic epidemiology and genomic research. In one international review of 43 meta-analyses describing 697 independent gene–disease association studies (Ioannidis, Ntzani, and Trikalinos 2004), for example, 76% (n = 224,546) of the individuals studied were classed as being of European descent (i.e., drawn from native populations of Europe or subjects of European descent from Oceania, North America, or South America, including Hispanics), 18% (n = 53,239) were of East Asian descent (i.e., from native populations of China, Japan, Korea, Indochina, and the Philippines), and only 3% (n = 7,961) were of African descent (i.e., African Americans or from populations of sub-Saharan Africa). Moreover, because the review included only meta-analyses that compared results from at least two "racial" groups, these numbers likely *underestimate* the degree to which European-descent samples have been the object of investigation in population-based association studies. The differential participation of non-European groups in genetic research is so taken for granted within the research community that these marked sample-size differences were not even commented on by the authors.

More recently, genome-wide association studies (GWAS) have come to prominence as a key method of genomic discovery. GWAS involve the simultaneous comparison of many thousands of common genetic variants scattered throughout the genome, between cases (individual participants who have the disease or trait of interest) and controls (participants who are matched with cases on many characteristics but do not share the condition of interest). Because the average effects of GWAS-detected genetic variants are quite small, typically many thousands of cases and controls must be compared to reliably identify a disease-associated variant. A recent review of 373 GWAS, as catalogued by the National Human Genome Research Institute, suggests that the underrepresentation of racial/ethnic minorities in such studies has been even more pronounced (Need and Goldstein 2009). As shown in Table 3-1, 96% of participants in single-population GWAS and 92% of participants in mixed (i.e., multipopulation) GWAS were found to be of European race/ethnicity. Not only have GWAS on individuals of European ancestry been performed at a ratio of nearly 10 to 1 versus all other groups combined, but average sample sizes (which affect the statistical power to detect genetic association) have been twice as large for European samples. As a result, racial/ethnic disparities in genomic research participation have become worse, not better, in the period since the completion of the Human Genome Project (Lander et al. 2001).

There are a variety of explanations for the significant underrepresentation of racial/ethnic minorities in genomic discovery science. To a degree, observed inequalities parallel differences that have been observed for other classes of biomedical research, including clinical trial research (Ford et al. 2008), and suggest major barriers to the recruitment and retention of minority research participants (James et al. 2008). Notorious human-subjects violations in the conduct of research with specific minority communities, such as the Tuskegee syphilis experiment (McCallum et al. 2006), are well known and widely discussed in communities, contributing to widespread distrust of biomedical research. Cultural, linguistic, and/or socioeconomic differences between academic researchers and minority communities can also substantially complicate recruitment efforts (Yancey, Ortega, and Kumanyika 2006). In addition, some researchers also acknowledge an analytical preference for populations of European ancestry, due to a perceived "greater ease of discovery" in such populations (Need and Goldstein 2009). Specifically, on average, Europeans have lower levels of within-population genetic variation and a higher degree of chromosomal association, or linkage disequilibrium, than those from non-European backgrounds, making the search for disease-associated variants more straightforward.

The contention that some populations are better suited for genomic discovery research than others reflects a growing recognition within the human genetic research community of the importance of population genetic variation. Despite long-standing consensus that there is "no biological basis to race" (Gould 1996; Graves 2004), numerous empirical investigations have affirmed that small, but statistically significant, differences in genetic variation exist among socially

Table 3-1. Ethnic Origin of Participants in Genome-Wide Association Studies

Race/ethnicity	Number of studies	Total participants		Percentage	Average sample size
European only	320	1,581,776		96%	4,943
Asian only	26	52,841		3%	2,032
Hispanic only	3	1,019		0.06%	340
Native American only	2	1,102		0.07%	551
Jewish only	2	3,479		0.2%	1,740
Gambian only	1	2,340		0.1%	2,340
Micronesian only	1	2,346		0.1%	2,346
TOTAL		1,644,903			
Mixed	11	European	92,437	92%	8,403
		African American	7,500	7.5%	682
		Asian	33	0.03%	3
		Papua-New Guinean	276	0.3%	276
		Other	269	0.3%	24
		TOTAL	100,515		

Adapted from Table 1 in Need AC, Goldstein DB. Next generation disparities in human genomics: concerns and remedies. *Trends Genet.* 2009;25(11):489–494.

constituted groups, particularly when those groups trace their heritage to ancestral populations that lived on different continents (Weiss and Fullerton 2005). Differences exist both in the level of genetic variation found within populations and in variation between groups. These differences are understood to have arisen as a function of regional demographic and selective forces; for example, regional differences in inherited blood disorders are believed to be related to the protective effects of such changes in regions of the world where malaria is endemic. The extent to which such differences explain disparities in health *outcome* across groups is unclear, and hotly debated (Frank 2007). Nevertheless, the unequivocal demonstration of population genetic differences has made the unequal participation of racial/ethnic minorities in genomic research an epistemic, as well as a political, concern. Recent calls within the genetic community for a more diverse basis for genomic discovery have been based, for example, in a recognition that current samples are not representative of U.S. populations (e.g., Collins and Manolio 2007) as well as a desire to "more comprehensively" identify genetic contributions to disease risk (McCarthy 2008), including ethnic-specific differences in disease risk (Tang 2006). However, the potential implications of sampling biases for later stages of the translational pathway—for example, efforts to design effective screening tests—have been, to date, relatively unexplored.

How worried should we be about systematic sampling biases in genetic and genomic research? There are at least three ways in which biases in genomic discovery can impact downstream genetic testing. First, a gene with an appreciable contribution to disease risk can be overlooked entirely if susceptibility variants in that gene occur at only a low relative frequency in the population chosen for discovery. That is, if a particular variant is uncommon in European-descended populations but relatively common in other groups, a study in which 99% of participants are European will likely fail to identify that variant as important. Second, a disease gene may be convincingly identified, but the full distribution of predisposing mutations within the gene may remain incompletely described. This can lead to the identification of "variants of unknown significance" (VUS), the nature of whose association with disease is uncertain. In this case, even if a validated genetic test is developed, individuals may receive results whose meaning or usefulness is unclear. Finally, a specific susceptibility variant may be replicably associated with disease, but the extent to which the variant increases risk may be inaccurately characterized or may indicate different relative risks in different discovery samples. Here, a test will detect a variant likely to increase disease risk but provide uncertain information regarding the degree of clinical intervention required. Examples of these scenarios are discussed below.

GENETIC DISCOVERY

One way in which over-focus on populations of a single racial/ethnic background can have an impact on discovery is by limiting attention to the genes (and variants within them) that occur at detectable frequency in that population. While this is not a problem if the same susceptibility factors are present to the same general degree in nonsampled groups, many common complex traits and diseases are determined by variants that do *not* occur in all populations to the same degree (McCarthy 2008). So, susceptibility variants common in non-European populations may be missed entirely in a study of European-based samples, and variants identified as important in samples from European descendants may contribute little to disease susceptibility in populations who trace their heritage to another region of the world.

Example

A 25-year-old man of Japanese descent has a history of diabetes on his mother's side of the family. He has put on 30 pounds since graduating from college. He is aware that obesity is a risk factor for type 2 (i.e., adult-onset) diabetes, and he is worried that the combination of his genetic makeup and weight gain may place him at increased risk. However, his father, who is also overweight, does not have diabetes, and he realizes that it is possible that he did not inherit his mother's (presumed) genetic susceptibility. He decides to have predictive genetic testing for type 2 diabetes risk, which he reasons will give him more specific information about his personal risk of developing the disease and hence help him decide how hard he should try to lose the excess pounds.

After researching potential alternatives, he opts for the Health Edition Test from 23andMe (www.23andme.com), a company that offers genetic testing services directly to consumers, without a physician order. The test, which costs $429, provides information about 156 diseases and conditions and includes a panel of 9 susceptibility variants that have been associated with diabetes susceptibility in European, Asian, and African populations.

When he receives his test results, he learns that while a few of the reported genotypes suggest a mildly elevated risk of developing type 2 diabetes compared with the general population, the composite conclusion from the test panel is that his risk of developing type 2 diabetes is in the "normal" range. He breathes a sigh of relief and decides he does not need to make substantial changes in his diet or join the gym; over time, he also neglects to see his doctor for routine checkups. In his early 40s, he begins to suffer adverse health effects resulting from undiagnosed diabetes.

What Went Wrong?

The test failed to provide accurate information to this individual because the gene(s) contributing to diabetes susceptibility in his family were not included as part of the genetic test panel. This can occur when the genetic variants contributing to disease risk are "private"—that is, unique to a particular family—but that is not what happened here. Instead, his results can be traced to the fact that the genetic test panel was based on data obtained predominantly from populations of European, not Asian, ancestry. There is at least one (Unoki et al. 2008; Yasuda et al. 2008) (and there possibly could be many more) susceptibility variant for risk of type 2 diabetes that is more common in Asian populations but relatively rare in Europeans, and hence not included in many testing panels. Only equivalently comprehensive investigation of genetic contributions to diabetes risk in a range of Asian racial/ethnic populations would have identified the gene as an important candidate for testing. The individual thus received a false-negative finding, which led to inappropriate reassurance about his potential for developing diabetes.

VARIANT DISCOVERY

Underrepresentation of non-European populations can also lead to an incomplete catalogue of the ways in which specific disruptions in a disease-associated gene contribute to disease risk. If a given gene demonstrates a limited number of possible disease-causing mutations (also called variants), and if the same causal mutations are present in all populations, then the study population used to identify those variants will not make a difference in the analysis. But if many hundreds, or potentially thousands, of different pathogenic mutations are in play, and if some fraction of these are confined to particular groups (McCarthy 2009), then a catalogue of mutations based on analysis of a patients from a single population will leave many causal variants unidentified.

Example

A 30-year-old African American woman whose mother had breast cancer and whose aunt had both breast and ovarian cancer is concerned about her risk of developing cancer. She seeks advice from a genetic counselor about testing for breast cancer susceptibility. After a thorough review of her family history, the counselor suggests that genetic testing could be informative with respect to her personal risk and arranges for genetic testing of the two known susceptibility genes for breast and ovarian cancer, *BRCA1* and *BRCA2*. (Testing in the United States is done at a single laboratory, Myriad Genetics, which holds the patents to the testing process.) If a specific mutation had been previously identified in a family member, or if she had any Ashkenazi Jewish ancestry, she could have a targeted test that would look for the particular variants seen in that group[1]; but as she is the first in her family to be tested and has no known Jewish ancestry, she will need the more comprehensive (and hence considerably more expensive) BRACAnalysis test (www.bracnow.com). This test involves complete sequencing of the *BRCA1* and *BRCA2* genes, rather than targeted genotyping of specific variants. Fortunately, because the woman's family history indicates that she may be at increased risk, her insurance will cover most of the cost.

After several anxious weeks, she receives a call from her counselor and goes in to receive her test results. There she is told that the test has identified a "genetic

1. Three common "founder" mutations (2 in the *BRCA1* gene and 1 in the *BRCA2* gene) are observed among many at-risk women with Ashkenazi Jewish ancestry (Warner et al. 1999). A panel which tests just these mutations is available for such women.

change of uncertain significance," meaning there is a mutation in one of the tested susceptibility genes that may affect protein function but has not been previously observed and so cannot be definitively said to increase cancer risk. Although the counselor had warned her that such a result was possible, she is upset to learn that her results will likely not be informative until the testing company has identified other women with the same mutation who have gone on to develop breast cancer. She has one sister, who does not want to undergo testing, and her other affected family members are all deceased. The counselor suggests that the best she can do, in the absence of additional information, is to undergo annual mammograms.

What Went Wrong?

In this case, while the test methodology can detect novel variants in known susceptibility genes (i.e., *BRCA1* and *BRCA2*), the clinical significance of the finding will remain unclear if the specific mutation identified has not been previously associated with presence of disease in other families. Such VUS are quite common in "comprehensive" (i.e., sequence-based) genetic tests. For VUS to be transformed into variants of confirmed clinical relevance, a sufficient number of women diagnosed with cancer must be found to share the same mutation. And while women of any ethnic background can receive inconclusive test results, there is evidence to suggest that minority women are disproportionately affected by such findings (Hall et al. 2009). In this case, if the woman's breast cancer mutation happens to be found more often among women of African ancestry, the fact that fewer such women have been previously tested means that it will be classed as "not seen before" even if, in fact, it is not a private mutation. This woman is disadvantaged because fewer women like her (and their families) have had their susceptibility genes examined for cancer-linked mutations.

EFFECT-SIZE ESTIMATION

Finally, even if a specific gene and its variants are well characterized and significantly associated with disease predisposition, biased sampling can interfere with the comprehensive investigation of the effect size of different variants. Effect size refers to the strength of the association with disease risk. As in the two previous examples, if the extent to which a specific genotype predicts disease outcome is consistent across populations, it makes no difference what population is used to estimate effect size: the results will apply to any individual, irrespective of racial/ethnic affiliation. If, however, background genetic factors or environmental exposures modify the strength of association such that effect size varies with population background (Bamshad 2005; Ioannidis, Ntzani, and Trikalinos 2004), estimates based on data from one or few discovery samples may not be applicable to individuals from other populations. In this case, a test will detect a variant likely to increase disease risk, but it will provide uncertain information regarding the degree of clinical intervention required.

Example

A 45-year-old Puerto Rican man with angina has been sent to a cardiologist for routine clinical follow-up. The cardiologist does a battery of tests but is not sure how aggressive a treatment plan to recommend. He asks whether cardiovascular disease runs in the family, but the patient does not have much information. He remembers that one uncle may have died "young," but he is uncertain of the cause, and he states that he is unable to consult with other family members to learn more. In the face of this uncertainty, the cardiologist recommends that the patient undergo genetic testing for known cardiovascular susceptibility variants.

The cardiologist suggests deCODEme's Cardio Scan test (www.decodeme.com/cardio-scan), which tests for the presence of susceptibility variants in a panel of genes. The test provides information about genetic risk for heart attack, abdominal aortic aneurysm, atrial fibrillation, peripheral arterial disease, intracranial aneurysm, and venous thromboembolism. When the test results are returned, the cardiologist reviews the findings with the patient. Risks for three of the six diseases examined vary by population ancestry and gender, while the other three risk estimates are based on results obtained from samples of European ancestry only; however, risk estimates for individuals of Puerto Rican ancestry are not available. The cardiologist is aware that people of Puerto Rican ethnicity may trace their ancestry to European and African forebears, and so he considers risk estimates relative to each population background. Unexpectedly, the patient's lifetime risk of heart attack is considerably higher if African rather than European ancestry is assumed. Even though the patient does not have any immediate known African heritage, the cardiologist recommends a surgical intervention out of "an abundance of caution."

WHAT WENT WRONG?

In this case, at least one of the susceptibility variants contributing to an increased risk of heart attack has been shown to affect risk differently in different populations. This is a well-recognized phenomenon in genetic epidemiological research, and one which is believed to reflect some systematic, but unmeasured, difference in either genetic background or environmental exposure that is correlated with ancestral background of a study sample (Bamshad 2005). Such population differences in estimated "effect size" severely complicate the interpretation of individual genetic risk, particularly when, as was the case for this patient, individuals understand themselves to be of mixed ancestral heritage. While current data suggest that population-level differences in effect size are uncommon (Ioannidis, Ntzani, and Trikalinos 2004), it is also true that, as was the case for the deCODEme panel, risk information is often available only for populations of European origin, which makes it hard to assess how often such differences actually arise. This individual was adversely affected both because his race/ethnicity is atypical of those for whom susceptibility information is usually generated and because he is not readily assignable to a definable ancestral heritage (such as those around which many genetic studies are constructed).

POTENTIAL REMEDIES FOR GENOMIC DISCOVERY BIAS

In each example described above, a failure to adequately investigate genetic contributions to disease risk in a particular population limits the clinical benefit of genetic testing for particular patients. On its face, this is a discouraging phenomenon, but recognizing the practical effect of biases in genomic discovery immediately suggests a potent remedy: eliminate or otherwise address population sampling biases. A fairer distribution of potential health benefit could follow such changes. (This thesis presupposes—somewhat optimistically—that we will be able to simultaneously overcome other barriers to healthcare access and delivery).

There are a few different ways in which systematic sampling biases in genomic research could be remedied as a matter of research policy. First, the *inclusion* of individuals from diverse ethnic groups, in numbers representative of national demographic proportions, could be required as a condition of funding. This sampling preference is, in fact, already a matter of U.S. federal research policy as per the conditions of the NIH Revitalization Act of 1993. Second, the *equitable*, as opposed to representative, ascertainment of individuals from multiple ethnic groups could be required, as recently recommended by Need and Goldstein (2009). Third, population-based ascertainment that takes known effects on statistical power into account (e.g., average levels of genetic variation), such that individuals from certain populations are *oversampled* relative to either representative or equity-based sampling schemes, might instead be encouraged. In the next section, each of these solutions is considered in respect to the effect they might have on the examples discussed above.

Representative Inclusion

As noted above, the inclusion of ethnic minority populations in biomedical research was required by congressional mandate as part of the 1993 NIH Revitalization Act. The NIH Inclusion Guidelines, which went into effect in 1994, require the following:

> [W]omen and members of minority groups and their subpopulations must be included in all NIH-supported biomedical and behavioral research projects involving human subjects, unless a clear and compelling rationale and justification establishes . . . that inclusion is inappropriate with respect to the health of the subjects or the purpose of the research (NIH 2001).

Investigators are asked to report and justify enrollment plans at the time of grant submission and to confirm participant demographics in study progress reports. Although this policy can be credited with increasing the participation of ethnic minority populations in genetic and other forms of biomedical research (Epstein 2007), it has done little to rectify marked imbalances in population sampling. This is due in part to the fact that the guidelines are enforced on a study-by-study basis, and population scientists can often justify an exclusive focus on one demographic group. Moreover, because of the way the policy has been implemented, in particular the required use of sociopolitical (i.e., defined by the Office of Management and Budget) categories of race and ethnicity for reporting, the policy has had the additional, unwelcome effect of promoting an analytical emphasis on biological differences among socially constituted groups, a phenomenon sociologist Steven Epstein has coined the "inclusion-and-difference paradigm" (Epstein 2007).

While prominent genetic epidemiologists and policy makers (Collins and Manolio 2007) have advocated that the involvement of research participants in numbers proportional to U.S. population demographics would represent an effective remedy to the sampling biases that threaten genomic translation, this is probably not the case. Because approximately 75% of the U.S. population currently self-identify as having European ancestry (Grieco and Cassidy 2001), a policy of representatively inclusive sampling would continue to support the disproportionate recruitment and analysis of individuals of European origin. In such a sampling regime, genetic discovery would likely continue to be pursued first, and most effectively (from the point of view of having sufficient sample size to achieve the required statistical power), in European-based samples. In the cases discussed above, the test for diabetes susceptibility would remain weighted toward the consideration of variants common in European samples but uncommon in other groups; breast cancer variants of uncertain significance would continue to disproportionately impact minority ethnicity test-takers; and where disease risk (as measured by the effect size of specific variants) is differentially distributed with respect to population background, individuals of European ethnicity would continue to benefit from better-validated and statistically more certain risk information. Hence, if equivalent benefit from genomic discovery is the desired endpoint, representative sampling will simply not suffice.

Equitable Sampling

Another suggestion, advanced in a recent commentary by Duke geneticists Need and Goldstein (2009), is to instead insist on the equitable, as opposed to merely representative, recruitment of minority and nonminority research participants. These authors contend that sampling biases have thus far not affected healthcare because GWAS have largely failed to identify clinically important genetic effects. But they warn that, with an anticipated shift to newly available whole-genome sequencing approaches that will identify many more clinically relevant variants, continuing inequities in genomic discovery could significantly exacerbate health care disparities in the future. To address this concern, they recommend a research policy that would require both the sequencing of equal numbers of samples from European Americans and African Americans (to level discovery efforts) and the collection of equivalently sized control populations (to provide a robust basis for distinguishing causal variants from background "neutral" genetic variation).

Because sequencing studies, at least in the near term, will involve far fewer participants than typical GWAS, Need and Goldstein reason that previously collected research samples would suffice for this purpose.

Leading genome scientists' prominent acknowledgement of population sampling inequities in the genome sciences, combined with concrete recommendations for their address, is laudable. Although prioritized investigation of previously collected samples would do little to remedy unequal rates of *participation* by individuals of minority ethnicity, a concerted push to equalize analyses could go much of the way toward addressing current discovery biases. Regarding the examples discussed above, independent but equivalent investigation of genetic contributions to diabetes risk in an East Asian or Asian American sample would have identified susceptibility genes too uncommon to be detected in Europeans. Similarly, with a larger pool of unaffected African American women to serve as controls, potentially many more pathogenic breast cancer mutations could be distinguished from background genetic variants.

One concern left unaddressed by these recommendations, relevant to the third case example, is the decision regarding which non-European populations to prioritize for analysis. Need and Goldstein (2009) endorse special sampling consideration for African American populations, who are not only underrepresented disproportionately in most biomedical research but are also widely regarded as particularly difficult to study from a genetic standpoint. For evolutionary reasons, individuals of African ancestry exhibit on average greater levels of genetic variation (Weiss and Fullerton 2005), which makes the search for genetic contributions to disease risk more arduous. But given that many other U.S.-based ethnic groups (such as Puerto Rican Americans) are equally poorly represented among current genetic research studies, relying on the analysis of previously collected samples risks neglecting investigation of these groups. At the same time, the equivalent analysis of *all* major ethnic groups may be neither feasible nor, in fact, strictly necessary: if the effect sizes of most genetic variants are not modulated by genetic ancestry or ethnic background, then exhaustive analysis is not required. However, in the absence of systematic consideration of genetic risk across diverse populations, we have no way of distinguishing variants whose effects vary in a population-specific manner from those whose effects do not. Analyzing samples from several (but not all) minority groups would help identify when more intensive investigation of specific genetic effects is needed.

Although equal sampling of individuals from multiple ethnic groups could reduce many of the biases expected to accompany whole-genome sequencing, it is not well suited to addressing all forms of genomic discovery bias. In particular, population-based association approaches, which rely on statistically distinguishing causal susceptibility variants from background genetic noise, can be compromised by high overall levels of variation and differences in variation among ancestral subgroups (Cooper, Tayo, and Xiaofeng 2008; Need and Goldstein 2009). In such cases, equivalent ascertainment may not result in equivalent opportunity for discovery and, by extension, for health outcomes benefit. As noted above, this problem is most pronounced for ethnic minority communities with large amounts of recent sub-Saharan African ancestry, as such populations generally have higher average levels of genetic variation, inherited in more complex chromosomal arrangements, than populations from other geographic regions (Campbell and Tishkoff 2008).

Oversampling

A policy of directed oversampling might hold the greatest potential for addressing discovery biases in certain cases. For some populations (e.g., those with substantial, recent sub-Saharan African ancestry), oversampling (i.e., studying people in numbers greater than their representative proportions) is needed to achieve comparable statistical power for discovery.

The most important point to emphasize is that sampling decisions can, and really should, be based in a consideration of the ultimate potential health benefit of the data generated, rather than other criteria aimed at ensuring "fair" (representative, equal, etc.) participation. In the case of the examples discussed above, it is likely that an Asian population sample of an equivalent size to the originally ascertained European sample would have been adequate for the identification of genes common to that group. It is also likely that equivalent estimation of effect size among Puerto Rican research participants would have been sufficient to address the concerns of the third example. However, the higher level of background genetic variation in samples of African origin suggests that some degree of oversampling of African American women with a positive family history of breast cancer would be needed to resolve an equivalent proportion of variants of uncertain significance in that ethnic group.

A number of important challenges would be involved in acting on such a policy recommendation. Sustained effort to increase the participation of African and African American communities in genetic and genomic research would be needed, including the use of innovative communication and recruitment strategies aimed at overcoming those communities' well-known and long-standing distrust of research and researchers (James et al. 2008). Such efforts must be accompanied by careful, scientifically informed justification for the required deviation from either representative or equitable sampling, which might otherwise be perceived (incorrectly) as inappropriately "privileging" or, alternatively, "burdening" African American participants. Further, to ensure that the effort involved in attracting greater numbers of participants to research is not wasted by subsequent analytical shortcomings, genome scientists would need to devote greater attention to refining methods for the detection and interpretation of complex genetic variation. Current genotyping tools, such as SNP chips, have been shown to perform less well in populations with greater degrees of African ancestry (Manolio, Brooks, and Collins 2008).

As this brief analysis demonstrates, if the translational *consequences* of specific classes of discovery bias are explicitly considered, it is relatively easy to distinguish and evaluate research policy alternatives. Such consideration not only suggests that current policy preferences (such as representative sampling) are objectively inadequate, but also helps explain why a one-size-fits-all approach to population sampling will not guarantee equal opportunity of translational benefit. The alternative—a mixed approach of equitable multiethnic analysis, augmented where appropriate by more intensive investigation of cross-population differences and targeted oversampling—holds greater promise and should be promoted by both federal and international research-funding agencies.

CONCLUSION

Although there are few data to bear out the claim, it seems likely that few bench-based genome scientists stop to consider the downstream translational implications of their research. This may be particularly true of scientists who interact rarely (if ever) with research participants. Faced with a defined empirical puzzle (e.g., "What gene or genes contribute to risk for this disease?," "Which variants in this gene explain disease risk?," or "To what degree is disease onset associated with inheritance of this variant?"), researchers may view the ethnic composition of the discovery sample as a simple pragmatic consideration rather than a moral issue. The best sample is the one that can be most easily obtained, on the shortest timetable, and that is that.

Yet, as illustrated here, population differences in genetic susceptibility to specific diseases, when combined with background differences in genetic ancestry among ethnic groups, do render the choice of study sample consequential. These consequences extend beyond purely scientific considerations, such as producing a skewed or incomplete detailing of genetic contributions to disease risk, an effect that has begun to be recognized and commented on by some genome scientists (McCarthy 2008; Tang 2006). Systematic biases in genomic discovery threaten to limit the translational promise of genomic information, denying potential benefits not only to a significant fraction of the U.S. population, but to the majority of individuals of non-European ancestry living around the world. More sustained, and nuanced, attention to the character of population sampling in discovery phase research must begin from the recognition of what is at stake: the just distribution of the benefits that emerge from human genomic research.

REFERENCES

Bamshad M. (2005). Genetic influences on health: does race matter? *JAMA*. 294: 937–946.

Campbell MC, Tishkoff SA. (2008). African genetic diversity: implications for human demographic history, modern human origins, and complex disease mapping. *Annu Rev Genomics Hum Genet*. 9:403–433.

Collins FS, Manolio TA (2007). Merging and emerging cohorts: necessary but not sufficient. *Nature*. 445:259.

Cooper RS, Tayo B, Xiaofeng Z. (2008). Genome-wide association studies: implications for multiethnic samples. *Hum Mol Genet.* 17(R2):R151–R155.

Epstein S. (2007). *Inclusion: The Politics of Difference in Medical Research.* Chicago, IL: The University of Chicago Press.

Ford JG, Howerton MW, Lai GY, et al. (2008). Barriers to recruiting underrepresented populations to cancer clinical trials: a systematic review. *Cancer.* 112(2):228–242.

Frank R. (2007). What to make of it? The (re)emergence of a biological conceptualization of race in health disparities research. *Soc Sci Med.* 64:1977–1983.

Gould,SJ. (1996). *The Mismeasure of Man.* New York, NY: W. W. Norton & Company.

Graves JL. (2004). *The Race Myth: Why We Pretend Race Exists in America.* New York, NY: Dutton.

Grieco EM, Cassidy RC. (2001). Overview of Race and Hispanic Origin. http://www.census.gov/prod/2001pubs/cenbr01-1.pdf. Updated March 2001. Accessed June 21, 2010.

Hall MJ, Reid JE, Burbidge LA, et al. (2009). BRCA1 and BRCA2 mutations in women of different ethnicities undergoing testing for hereditary breast-ovarian cancer. *Cancer.* 115(10):2222–2233.

Ioannidis JP, Ntzani EE, Trikalinos TA. (2004). "Racial" differences in genetic effects for complex diseases. *Nat Genet.* 36(12):1312–1318.

James RD, Yu JH, Henrikson NB, Bowen DJ, Fullerton SM; Health Disparities Working Group. (2008). Strategies and stakeholders: minority recruitment in cancer genetics research. *Community Genet.* 11(4):241–249.

Lander ES, Linton LM, Birren B, et al. (2001). Initial sequencing and analysis of the human genome. *Nature.* 409(6822):860–921.

National Institutes of Health. (2001). NIH Policy and Guidelines on The Inclusion of Women and Minorities as Subjects in Clinical Research – Amended, October, 2001. http://grants.nih.gov/grants/funding/women_min/guidelines_amended_10_2001.htm. Accessed September 20, 2010.

Manolio TA, Brooks LD, Collins FS. (2008). A HapMap harvest of insights into the genetics of common disease. *J Clin Invest.* 118(5):1590–1605.

McCallum JM, Arekere DM, Green BL, Katz RV, Rivers BM. (2006). Awareness and knowledge of the U.S. Public Health Service syphilis study at Tuskegee: implications for biomedical research. *J Health Care Poor Underserved.* 17(4):716–733.

McCarthy MI. (2008). Casting a wider net for diabetes susceptibility genes. *Nat Genet.* 40(9):1039–1040.

McCarthy MI. (2009). Exploring the unknown: assumptions about allelic architecture and strategies for susceptibility variant discovery. *Genome Med.* 1(7):66.

Need AC, Goldstein DB. (2009). Next generation disparities in human genomics: concerns and remedies. *Trends Genet.* 25(11):489–494.

NIH Revitalization Act of 1993 (PL 103–143), 42 USC Sec.289a-1 (1993).

Tang H. (2006). Confronting ethnicity-specific disease risk. *Nat Genet.* 38:13–15.

Unoki,H, Takahashi A, Kawaguchi T, et al. (2008). SNPs in KCNQ1 are associated with susceptibility to type 2 diabetes in East Asian and European populations. *Nat Genet.* 40(9):1098–1102.

Warner, E, Foulkes, W, Goodwin, P, et al. (1999). Prevalence and penetrance of *BRCA1* and *BRCA2* gene mutations in unselected Ashkenazi Jewish women with breast cancer. *J Natl Cancer Inst.* 91:1241–1247.

Weiss KM, Fullerton SM. (2005). Racing around, getting nowhere. *Evol Anthropol.* 14:165–169.

Yancey AK, Ortega AN, Kumanyika SK. (2006). Effective recruitment and retention of minority research participants. *Annu Rev Public Health.* 27:1–28.

Yasuda K, Miyake K, Horikawa Y, et al. (2008). Variants in KCNQ1 are associated with susceptibility to type 2 diabetes mellitus. *Nat Genet.* 40(9):1092–1097.

4

The Autism Genetic Resource Exchange: Changing Pace, Priorities, and Roles in Discovery Science

HOLLY K. TABOR AND MARTINE D. LAPPÉ

Human genetics research depends on the availability of DNA samples and data from people, and often families, with a given disease or condition. Historically, genetics researchers focused their recruitment on families affected by relatively rare diseases. These individuals were usually identified at genetics clinics at academic medical centers, where clinician-researchers had close relationships with affected families and the larger disease communities. In these settings, families were often identified as potential research participants through clinical activities including diagnosis, reproductive counseling, and general patient care.

This traditional paradigm of genetic research design and recruitment has been challenged by the increasing focus of genetic research on more common, complex diseases such as heart disease, diabetes, and autism. In conditions like these, the contribution of any single genetic change to disease risk may be small. As a result, large numbers of participants are required in order to detect association between a genetic variant and disease risk. This need for large-scale study recruitment has challenged traditional models of genetic research recruitment, in part because researchers seldom have the same kind of established clinical relationships with complex disease populations as they have with rare disease communities. Even when researchers do have strong connections to affected individuals, the potential local recruitment is often too small to meet the statistical requirements for association studies.

Theoretically, the need to acquire sufficiently large study populations could be met by researchers pooling their data from smaller studies or local recruitment efforts. However, the competitive environment within academic science often discourages this kind of sharing, as researchers need to lead studies in order to obtain research funding, publish manuscripts, and get tenure at universities. This atmosphere leads to a situation where many small studies proceed slowly, as each lacks the ability to adequately recruit subjects to complete research analyses.

Autism is an example of a common disease for which this traditional genetic research paradigm has been challenged. This chapter focuses on the history of the Autism Genetic Resource Exchange (AGRE). In 1997, AGRE was created by a parent advocacy organization called Cure Autism Now (CAN). The parents who founded CAN and AGRE sought to dramatically change the way that basic genetic research on autism was conducted, prioritizing the acceleration of research and translation. Leveraging researchers' need for large data sets, AGRE's creators advocated for a shift away from the traditional genetic research model of investigator-driven recruitment.

These parent advocates developed a new research model that shifted data and sample collection out of the exclusive hands of researchers and into the domain of an advocacy group. AGRE made recruitment and management of the data resource their "core competency" by recruiting families directly and creating a large, openly available resource of samples and phenotypic data. This open resource changed the kind of genetic research on autism that was possible.

This chapter explores the story of AGRE, its impact and limitations, and its implications for future models of genetic discovery science. The first section provides important context for the story of AGRE by describing changing approaches to understanding the causes of autism in the twentieth century. The second section relays the story of AGRE, focusing on the efforts of two parent advocates, Jonathan Shestack and Portia Iverson. The third section explains the impact of AGRE on the discovery stage of autism research. Finally, the fourth section considers some of the challenges that have faced AGRE, and how they translate to the potential limitations of using AGRE as a model for transforming discovery-stage research in other disease contexts.

FRAMING AUTISM CAUSATION IN THE TWENTIETH CENTURY: FROM REFRIGERATOR MOTHERS TO GENETICS

To understand some of the challenges of autism genetic research, it is important to understand the history of the disorder and research into its causes. Autism is understood to be both a psychiatric diagnosis and a neurodevelopmental disorder, characterized by impairments in socialization, abnormal verbal and nonverbal communication; and restricted, stereotyped behaviors and interests (APA 1994). These symptoms usually become apparent in the first few years of life but may be quite heterogeneous across cases, with substantial variability in verbal and communication ability, behavioral challenges, and co-occurring health conditions. Research efforts related to autism have been influenced by different disciplinary, technological, social, and political changes over time.

Soon after the psychiatrist Leo Kanner first described autism, in 1943, other experts began to suggest that autism was caused by psychological influences in the child's environment. These included parental attitudes, bad parenting, and cold mothering, or what has become known as the "refrigerator mother" hypothesis of autism causation (Orsini 2009). In his 1967 book, the psychologist Bruno Bettelheim described and popularized this theory, writing, "The precipitating factor in infantile autism is the parent's wish that his child should not exist" (Bettelheim 1967, p. 125). This view became accepted among clinicians and prevalent in society.

In the late 1960s and 1970s, understandings about causes of autism shifted. Bernard Rimland, a psychologist and father of a child with autism, worked diligently to dispel the idea that "refrigerator mothers" caused autism and suggested instead that autism was biological in origin (Silverman 2004). He pioneered a movement of parents advocating for the understanding that autism was a biological and medical disease, rather than a psychological condition. This movement laid the groundwork for research-focused advocacy, suggesting that scientific research to identify the causes of autism could and would provide options for treatment and prevention.

In the same period, researchers began to apply genetic research tools, such as twin and family studies, to estimate the heritability or genetic component of a wide range of diseases and traits, including autism (Muhle, Trentacoste, and Rapin 2004). In 1977, the psychiatrist Michael Rutter, together with psychiatrist and researcher Susan Folstein, conducted an autism heritability study in 21 twin pairs. Their results demonstrated that the chance of siblings of children with autism being affected was between 2% and 7% (Folstein and Rutter 1977). These results represented a 50- to 100-fold increase over the known population risk for autism. This study also indicated that identical twins were more likely to both be autistic, or concordant, than fraternal twins, suggesting that shared genetic factors played an important role in the disease (Szatmari and Jones 2007).

As a result of these and similar studies, the scientific and medical community began to understand autism not as a psychosocial disorder but as a genetic disease. Soon after the Folstein and Rutter paper was published, scientists and clinicians at autism research centers in the United States and United Kingdom began recruiting families with children diagnosed with autism to participate in large genetic family studies called *linkage studies*. This research approach was modeled after the strategy that had been successful for single-gene diseases with simple models of Mendelian inheritance, such as cystic fibrosis and Huntington's disease. Researchers hoped that genetic linkage studies would help identify the gene or genes that caused autism. These studies were facilitated by the development of new diagnostic tools: the Autism Diagnosis Interview (ADI) and the Autism Diagnostic Observation Schedule (ADOS). These tools made autism diagnosis and classification more systematic and consistent and facilitated the standardization of autism clinical research.

While linkage studies seemed to confirm an overall contribution of genetics to autism causation, they failed to consistently identify specific causal genes (Szatmari and Jones 2007). One study suggested that at least 10, and possibly as many as 20, genes might contribute to autism etiology (Risch et al. 1999). These studies made it clear that autism—unlike classic Mendelian genetic diseases—was not caused by changes in a single gene, or even in a small number of genes, acting in a simple dominant or recessive manner. Rather, autism was framed as a "complex disorder" with many genes of small effect and variable models of inheritance. It was against this backdrop of limited genetics research with frustrating outcomes that the story of AGRE unfolded.

TWO FRUSTRATED PARENTS AND A SICK CHILD

The history of AGRE begins with the experience of one family. In 1995, Jonathan Shestack and Portia Iversen's three-year-old son Dov was diagnosed with autism. The story of their journey from this point has been well documented in the media (Bazell 2005) and in Iversen's own book, *Strange Son* (Iversen 2006). Shestack and Iversen's first action was to identify doctors and treatments to help their son. Like many parents of children with autism, they were told that there were few, if any, medical treatments or interventions for autism.

As a result, Shestack and Iversen actively sought out research studies that they could participate in, or possibly learn from, to help Dov. On searching the literature and communicating with scientists, they were frustrated to discover that, from their perspective, very little autism research was being conducted. As Iversen searched medical libraries for published journal articles about autism, she was dismayed by the lack of scientific research and knowledge about the disease. She framed it as an issue of justice, saying:

> I think a lot of people just assumed that diseases have some sort of built-in equality. If a disease is really bad, there's someone researching it, and if it affects a lot of people, there's more people researching it . . . I just had this kind of very naïve concept. So it was a shock when we realized that there wasn't any research. I mean, almost none. (Iversen interview, April 17, 2009, Los Angeles, CA)

The family continued their search for medical treatment and met with clinicians, who invariably told them that there was no medical treatment they could offer to help their child. Iversen found this claim illogical, given the absence of research in the field, and it fueled her motivation to change the scientific and medical approach to autism:

> It just outraged me that there was absolutely nothing known, and yet it was being called incurable . . . that just really revved me up and got me mad! Because this is my kid, right? He's incredibly sick, no one's done due diligence about why, and I'm being told not to do a thing because it can't be fixed.

That is just a paradigm I can't live with. (Iversen interview, April 17, 2009, Los Angeles, CA)

BECOMING BIOMEDICAL ADVOCATES

The unwillingness to accept this paradigm spurred Iversen to become what she called a "biomedical advocate," or an advocate for biomedical research. She explained: "This is an illness. We rely on science for cures to illnesses. We have to go to science, but we can't go blindly" (Iverson interview). Shestack and Iversen both worked in Hollywood, Shestack as a movie producer and Iversen as an Emmy award–winning art director in television. Drawing on the family's professional experiences in film, Iversen shared the following analogy:

> It's like when you shoot a movie. If you're a director, you may not have to know how to run the camera. But if your cinematographer tells you that the big scene that you must have is impossible, you have to know enough to know whether that's true or not. You may not know everything about it, but you have to know enough—because you're going to get the lazy cameraman, you're going to get the disinterested, you're going to get the would-be director who wants to do it their way. Same with scientists. They all have their own motivations and different agendas, because this is their career. They don't have a kid with the disease, usually. So you have to know enough so that you can determine whether what they're telling you is reasonable. (Iversen interview, April 17, 2009, Los Angeles, CA) They believed that parent advocates needed to become knowledgeable about the science and learn how to communicate directly with scientists in order to influence autism research priorities.

Iversen and Shestack decided to start a new advocacy organization in which parents would advocate for research by being active partners in the research enterprise itself, rather than simply a resource for recruiting research participants or raising money. Iversen said:

> So we understood, "OK, we have to have this research foundation. We don't want to be passive fundraisers who give money to science, and then just hope they do something, since we know that they are far less motivated than we are, because we wake up with an autistic child in our house, and they don't." (Iversen interview, April 17, 2009, Los Angeles, CA)

They called their organization Cure Autism Now, or CAN. The name was intended to convey the sense of urgency and the focus on the translation that they felt should accompany autism research. Geraldine Dawson, the current Chief Officer of the advocacy organization Autism Speaks, described the impact of CAN on the field at the time: "The urgency, the idea of a cure—no one said that word.

They just didn't see it. And so it was completely a stretch goal that I think motivated the field and the hope [of parents]..." (Dawson interview, October 10, 2007, Seattle, WA). Through the establishment of CAN, Iverson and Shestack were able to convey their belief, and that of other parents, that autism could in fact be curable, and therefore was a worthy target for intense scientific efforts.

The couple's professional background in the movie industry, as well as their West Coast location, influenced their decision to focus specifically on autism research. Shestack put it this way:

> We didn't know anything about [starting a research foundation], really, but felt, sort of naively, that because . . . we knew politicians, and I knew movie stars, it wouldn't be so hard. And I honestly believe that living in California was a big part of that. I think . . . if I had been still on the East Coast, under sort of the oppressive sway of "the Academy [i.e., academia]," it would never have happened. And, in fact, anybody we ever met from that ilk would try and talk us out of it. (Shestack interview, April 17, 2009, Los Angeles, CA)

This perspective led Shestack and Iverson to challenge the conventional wisdom about what was and was not possible in scientific research.

CHALLENGING ASSUMPTIONS ABOUT THE PACE OF SCIENCE AND COLLABORATION

One of CAN's key principles was a rejection of the claim that science cannot be hurried. When Shestack and Iversen met with researchers who were conducting genetic research on autism, they were told that it took many years, perhaps decades, to recruit enough families to conduct research. Science could not be hurried, the researchers explained, and parents should be patient and allow researchers to focus on doing the research the way it should be done. Shestack and Iversen refused to accept this position. As Shestack said later, "I work in the movie business . . . I know you can hurry anything. You put more guys on the job, you spend more money. You just can't do it for free" (Coukell 2006, p. 28). With the investment of enough personnel and resources, he and Iversen believed that scientific research, like film production, could in fact be pushed to move more quickly.[1]

In addition to researchers' belief that science could not be hurried, Iversen and Shestack identified another problem: lack of collaboration and sharing of essential

1. In starting a research-focused advocacy foundation, Shestack and Iversen modeled their efforts on the examples of parent and family advocates for research on other diseases, such as the Juvenile Diabetes Research Foundation (JDRF 2010) and the Elizabeth Glaser Pediatric AIDS Foundation (Iversen Interview and Shestack Interview, April 17, 2009, Los Angeles, CA). By focusing primarily on genetic research, they followed the example of advocacy groups for Mendelian diseases, such as the Hereditary Disease Foundation's efforts in Huntington's disease (Wexler 1996).

scientific resources in the autism research community. Early in the process of establishing CAN, Shestack and Iversen met with leading genetics researchers in autism and asked them to make their samples and data available to other scientists. By sharing samples, they argued, the science would progress faster than if each researcher were working alone (Coukell 2006; Iversen interview; Shestack interview). The response of most established autism genetics researchers, according to Shestack, was to refuse to share samples: "All these people had just said no. In fact, they were very insulting about it and sort of said, '*Who* are you? Why are you doing this? You don't know anything.' . . . I have to say that the field did not distinguish itself by its generosity or maturity" (Shestack interview). Shestack explained this refusal to share samples:

> There are reasons. They're not understandable reasons, but it's hard to recruit families. If you [could] get them, it was expensive to process them . . . . You're in a constant battle with your colleagues and competitors to publish, to patent . . . . And these families are your currency. (Shestack interview, April 17, 2009, Los Angeles, CA)

Shestack, and Iversen consulted with several researchers who supported CAN. These researchers advised them that the advocacy group should "become the data." As Iversen described the conversation, "They said, 'The problem is, you need samples.' So the best thing [CAN] could do is start a gene bank and repository and set up a system that will allow everybody to work on the disease . . . . The best way is to create a resource, and then people will come and use the resource" (Iversen interview, April 17, 2009, Los Angeles, CA).

This strategy meant that CAN could control the resources, instill the organization with values that were important to the families involved, and remove researchers from decision making about data access and sharing. Shestack said, "It was *silly* to try and collaborate and change them [the researchers] . . . . What you had to do is just become the data, yourself, and just control it. And, if you cared about universal access, then you just had to have universal access through your own machine" (Shestack interview, April 17, 2009, Los Angeles, CA).

AGRE: THE 500-POUND GORILLA

In 1997, CAN founded AGRE specifically to pursue this goal of "becoming the data." AGRE would send staff around the country to conduct visits in the homes of families with more than one child affected by autism, gathering phenotypic data as well as tissue samples from which DNA could be extracted. AGRE would be an international resource: any researcher anywhere could apply for free access to the data contained in the AGRE database. For a small fee that was less than the costs associated with recruitment and sample preparation, researchers could obtain a DNA sample or cell line from the biobank. As a condition of access to AGRE resources, researchers were required to agree to deposit any data they generated in an open AGRE database that would in turn be made available to other approved AGRE users.

This strategy exploited the lack of success researchers were experiencing in identifying genes for autism using the "old" (i.e., small-cohort, single-researcher) paradigm of genetic research. Daniel Geschwind, at the time a neurologist and assistant professor at the University of California, Los Angeles, was recruited by Shestack and Iversen to chair the AGRE Steering Committee and serve as the Chief Scientific Advisor.[2] Geschwind believed that in order to advance the state of knowledge about autism, scientists needed to recruit significantly larger numbers of affected families into genetic research. Remembering his perspective at the time, he said, "My thought was, we're going to need at least 500 to 1,000 families to start this procedure, you know, to really have any power. People are talking about doing genome scans with 70 and 100 families, and that's just not going to do it" (Geschwind interview, May 20, 2009, Los Angeles, CA). Because AGRE could deliver the necessary samples and data, it could also dictate the terms of access and mandate broad data sharing. Clara LaJonchere, who became the AGRE Program Director in 2003,[3] described the approach:

> You [AGRE] take family recruitment and data collection completely out of the hands of researchers and you make it your core competency. You add several layers of quality control to that, but you turn around and make it an open resource. Researchers will come because, in order to do a full-scale autism study . . . it costs a lot of money and it takes a lot of time. (LaJonchere interview, October 10, 2008, phone interview)

Provided that researchers trusted the quality of the data, they would eventually engage with AGRE if they hoped to remain competitive in the field. Geschwind said, "AGRE totally changed the landscape of autism genetics . . . [it became] the 500-pound—or 1,000-pound—gorilla." As researchers realized "that they couldn't do it and claim the credit for it themselves," they became willing to use and work with the resource (Geschwind interview, May 20, 2009, Los Angeles, CA).

THE IMPACT OF AGRE ON THE DISCOVERY STAGE OF AUTISM RESEARCH

The story of the founding of AGRE by CAN is an example of shifting control of the discovery stage of the translational cycle of research. The parents involved

2. Geschwind is currently Professor of Neurology and Psychiatry and Behavioral Sciences and Director of the Center for Autism Research and Treatment at UCLA. He is also a member of the AGRE Steering Committee and AGRE Chief Scientific Officer.

3. LaJonchere is currently the Vice President of Clinical Programs at Autism Speaks. CAN merged with the advocacy organization Autism Speaks in 2007, and Autism Speaks now oversees AGRE.

in CAN decided to become "biomedical advocates," fluent enough in the language of science to engage with the research community. They recognized the absence of research on autism, specifically genetic research, and sought to focus their efforts on filling this gap. Rather than being passive fundraisers, they sought to work with researchers to address questions about sample size, recruitment, data sharing, and open access, as well as the timeline of research (Iversen interview, April 17, 2009, Los Angeles, CA).

CAN insisted, however, that discovery-stage research, including genetic research, explicitly focus on the ultimate goals of translation, and on reducing the time to achieve them. This approach challenged the existing paradigm that genetic research was often slow and usually focused on understanding biology and etiology, rather than developing, or even anticipating, cures and treatments. This accelerated research timeline was driven by the sense of urgency that parents felt in caring for their children with autism every day, and the desperate and pressing need to find translational outcomes of research that could help these children and help them soon.

Since its founding, AGRE has achieved its goals of establishing a large, open, and shared research resource. Between 1997, when AGRE was founded, and 2009, AGRE recruited more than 1,000 families from around the country (AGRE 2010), creating one of the largest resources for autism, and perhaps the largest openly available resource for any disease. For each of the families recruited, AGRE researchers collected diagnostic and phenotypic data as well as blood samples, and they prepared DNA and cell lines to be made available to researchers. Over time, AGRE made linkage and genome-wide association study data available for analysis. They also collaborated with researchers to recruit AGRE participants into ancillary research studies related to autism (LaJonchere, oral communication, July 16, 2010).

In the same period (1997 through 2009), 250 researchers from more than 22 countries were granted access to AGRE data (LaJonchere communication). This broad access suggests that AGRE has been successful in making genetic data on autism families available to a large number of researchers. These researchers have been productive, with more than 160 scientific publications using AGRE data released between 2001 and 2009. These publications demonstrate a substantial increase in research on autism, due at least in part to the availability of the AGRE resource.

AGRE has also served as a model for new studies funded by parent advocacy groups. Many genetic studies require families to have multiple siblings affected by autism as a condition of enrollment. Jim and Marilyn Simons were frustrated with the lack of opportunities for families like theirs, with only one child affected by autism, to participate in autism genetic research studies (Regalado 2005). In 2005, their foundation, the Simons Family Foundation, initiated a nationwide study called the Simons Simplex Collection Project, which is enrolling 2,000 families with only one child with autism. While there are important differences between the AGRE approach and that of the Simons project, there are strong similarities in the role of parents in collaboration with key research allies.

Since AGRE's founding in 1997, there have also been developments and changes in the broader scientific community in terms of the way research should be done, particularly with respect to data sharing and parent/patient involvement in research. In 2007, the National Institutes of Health (NIH) required all investigators receiving funding for genome-wide association studies to deposit their data in an NIH-run data repository, the database of Genotypes and Phenotypes (dbGaP) (Mailman et al. 2007). This national data repository was created in part as a response to some of the same challenges of complex diseases as those that faced autism. Large sample sizes, perhaps across multiple study cohorts, would be required to have statistical power to identify genes for these traits. It was also an effort to adopt some of the lessons learned from the Human Genome Project (Kaye et al. 2009), in which all genetic sequence data was rapidly put into the public domain in order to accelerate scientific progress.

There have also been new perspectives in biomedical research concerned more broadly with the kinds of roles that patients and research participants should play in making decisions about the research agenda and data access (Kaye et al. 2009; McGuire et al. 2008; Trinidad et al. 2010). The amount of participant involvement and the form that it takes has varied, from participant advisory groups to the informed cohort model (Kohane et al. 2007; Winickoff and Winickoff 2003). There have also been surveys of the general public and patient populations suggesting that participants want to be informed about possible uses of their samples and data, and in some cases would prefer to be asked permission before samples are used for new purposes (Goldenberg et al. 2009; Hull et al. 2008; Kaufman et al. 2009). Concern about the use of newborn screening blood spots for unauthorized research has led to advocacy efforts and, in some states, legislation requiring either explicit parental consent for use or destruction of the samples (Neergaard 2010).

These developments in data sharing and family/participant involvement have occurred in parallel with the efforts and successes of AGRE as a model for an advocacy-group-run biobank resource. While they cannot be attributed to AGRE, they reflect broader contextual changes in the landscape of biomedical research that are important for understanding the impact, implications, and limitations of organizations like AGRE.

POTENTIAL LIMITATIONS OF AGRE AS A MODEL FOR OTHER DISEASE CONTEXTS

While AGRE has achieved many successes, it also has faced several challenges, which translate into several potential limitations in its use as a model for how parent or patient advocacy groups can transform discovery research. First and perhaps most importantly, despite its successes in recruiting families to participate, attracting researchers to use the data, and increasing the number of scientific publications on autism, AGRE has not yet reached its goal of finding a cure for autism. While great progress has been made, and some genetic factors have been identified that account for a small proportion of the population risk for autism

(Buxbaum 2009; Durand et al. 2007; Pinto et al. 2010) neither AGRE nor any other research efforts in autism to date have successfully identified genetic factors that explain the causes of most cases of autism. These limitations in the discovery stage of the translational cycle have in turn inhibited the development of the treatments or cures that families had hoped for.

This situation presents a challenge for AGRE and for similar research-focused parent advocacy groups: By setting a stretch goal that motivates families and scientists, advocacy organizations may unintentionally create unrealistic expectations and false hope. As a result, families may lose faith and trust in the research enterprise. Perhaps more concerning, participants may experience therapeutic misconception, or the incorrect belief that participation in research is likely to directly benefit them or their child. The challenge for stakeholders who embrace this approach is to balance a focus on speed and translation with the setting of realistic expectations and education about the inherent challenges in biomedical research.

Funding represents a second potential limitation of this model. AGRE was originally funded entirely by its parent advocacy organization, CAN. The funding came from the efforts of parents nationwide to raise funds through grassroots efforts, such as fundraising walks, that also raised awareness of the condition and the need for more research (Solovitch 2000). CAN also sponsored more traditional fundraisers for major donors and used celebrity endorsements, particularly through Shestack and Iversen's connections within the film industry. Ultimately, however, this has not proved to be sustainable. In 2001, AGRE applied for a grant from the National Institutes of Mental Health (NIMH), and CAN and NIMH engaged in a partnership to fund and manage the AGRE resource. This collaboration amounted to recognition by NIMH of the value of the AGRE resource for the scientific community. However, it also reflects a recognition by CAN, the families, and the scientists involved in AGRE, that it was not possible for them to continue to be self-supporting: too much money was needed. They also did not believe it was fair to expect families to shoulder this burden. Speaking about the NIMH agreement, Geschwind said:

> We believed that such a resource should not only be on the shoulders of the parents, that the National Institutes of Health should be funding this. Such a public-private partnership would be beneficial in every disease . . . . And, because it was our baby . . . that it was really critical that we maintained some control over it, to insure that its core mission and high quality would remain true. But, also we wanted NIH to be a partner. And we felt that it was perhaps not fair to make the parents pay $1.5 million a year for something that was so critical to research. (Geschwind interview, May 20, 2009, Los Angeles, CA)

There is an important balance between funding and control in the research enterprise. By funding AGRE themselves, CAN and the researchers they worked with set the agenda, framework, and rules of their discovery research in a way that might not otherwise have been possible. Later, by partnering with NIMH/NIH after demonstrating early success with this approach, they compromised in part on control of the resource. They agreed to more traditional measures of objectives and deliverables in exchange for greater and possibly more sustainable resources. NIMH/NIH also compromised in agreeing to a more engaged role of the advocacy organization in conducting research and managing discovery resources.

This history leads to important questions for the discovery stage of research. What is the best way to manage this balance between funding and control? Who should be responsible for making sure that families, patients, and communities are involved in setting research priorities, and how? Should funding and control be completely linked, or should other governance or consultation and engagement structures be used to share control with stakeholders? These are important issues and may vary across disease and research contexts, as well as across the landscape of different kinds of advocacy organizations.

A third limitation of AGRE as a model is the uniqueness of its founders. CAN and AGRE resulted from the dedicated involvement of two exceptionally resourceful, persistent, and persuasive parent advocates, Jon Shestack and Portia Iversen. They were well connected and willing to reach out to and talk with anyone and everyone they thought could help their cause; they were relentless in pursuit of their goals. Without their involvement, AGRE might never have been founded or, once founded, might not have been successful. There are many active and successful advocates for biomedical research in other disease and population communities, and reaching out to legislators has become a more common part of disease advocacy. However, not all communities are likely to have advocates with the dynamic personalities, resources, and connections of Shestack and Iversen. Without other mechanisms for the recruitment and training of disease advocates, as well as incentives for researchers to partner with them at the discovery stage, it may not be possible for parents and/or patients to create resources like AGRE.

A fourth limitation of AGRE as a model is in the area of governance. AGRE was never designed to represent the interests of all parents of children with autism, or the interests of adults with autism. It was originally focused on biomedical research, specifically on genetics, so it is possible that families with different perspectives on the disease and its causation have been less likely to participate.[4] In the case of autism, not all advocates and stakeholders have agreed that genetic research, or even medical research, should be the highest priority. For example, more recently, those who espouse the movement for neurodiversity argue that the behaviors such as those seen with autism lie within the range of normal human behavior. These advocates believe that autism is not a condition that needs a cure or treatment, and that genetic research on autism is tantamount to eugenics (abfh 2005; Chew 2009; Kalb 2009; Taylor 2007). Other groups, such as Generation Rescue, believe that environmental causes play a critical role in autism and that genetic research disproportionately diverts research funds and effort away from

4. In recent years, AGRE has included research on environmental risk factors for autism, partly in response to requests from both researchers and parents (LaJonchere communication).

what they believe is most important (Belli 2010; Greenfeld 2010; Kirby 2009). The ongoing debates within and across autism advocacy groups reflect the heterogeneity of advocacy efforts surrounding autism, and in this landscape AGRE could not, and did not, represent all advocacy stakeholders.

AGRE consciously focused on the experience of the families that did participate in the resource, in part because it was founded by parents of children with autism. Participant engagement, recruitment, and data collection practices were tailored to reflect the needs and interests of these parents, including a regular research newsletter for enrolled families called *Listening to You* (LaJonchere interview). However, AGRE was not designed with a representative governance structure. Therefore, it is possible that some families' opinions and perspectives were not known to study coordinators and were therefore not considered in making decisions about how to manage the database. These might include perspectives on what research to prioritize, who should have access to the data, or what results should be returned to participants.

It is not clear whether or how a more representative model of governance would have affected the success and impact of AGRE, or whether participants would have wanted such an approach. AGRE may be instructive as an example of a nonrepresentative governance structure that can be successful for biobanking and the advancement of discovery research. By taking on a model of respect, treating families as an asset, and maintaining transparency and open communication, organizations like CAN and AGRE, as well as researchers in traditional research recruitment, can potentially involve a larger group of families in active research participation and engagement. Increasing awareness of research and how it is done, rather than engaging large numbers of families in direct-governance decision making, may prove to be an effective approach for some advocacy groups building discovery-stage resources in other disease contexts.

CONCLUSION

As genetics continues to evolve as a field, there will be new challenges in study design, recruitment, data sharing, and translation. The story of AGRE is an example of the efforts of a small number of committed parent advocates changing the discovery stage of genetics research. AGRE was successful because it filled an unmet need of researchers and the field, and because large numbers of families with autism were willing to support and participate in the AGRE model of an advocacy-group-run resource. There are several other models of advocacy involvement in discovery-stage research (Crenson 2008; Sharp et al. 2008; Terry et al. 2007). It is likely that no single model will work in every disease or research context, and there may be advantages to having several models in operation even within the same disease area. By studying AGRE, stakeholders in genetic research can learn how to anticipate some of the challenges of genetic research on complex diseases and develop innovative solutions that can change the pace, priorities, and roles in discovery research in the future.

ACKNOWLEDGMENT

The authors would like to thank the participants in our interviews for sharing their time and experiences.

REFERENCES

abfh. (2005). Commercial eugenics. Whose Planet Is It Anyway? [blog]. http://autisticbfh.tripod.com/eugenics.html. Updated February 21, 2006 [originally posted July 2005]. Accessed November 19, 2010.

[AGRE] Autism Genetic Research Exchange. (2010). Autism Genetic Research Exchange Web site. http://www.agre.org/. Accessed June 25, 2010.

[APA] American Psychiatric Association. (1994). *Diagnostic and Statistical Manual of Mental Disorders IV*. Washington, DC: American Psychiatric Association.

Bazell R. (2005). Parents push for autism cure: doctors credit parents for making research a priority. MSNBC Web site. http://www.msnbc.msn.com/id/7012176/ns/nightly_news. Updated February 23, 2005. Accessed June 25, 2010.

Belli B. (2010). The search for autism's missing piece: autism research turns its focus to environmental toxicity. *EMagazine*. 21(1). http://www.emagazine.com/view/?4984. Accessed June 25, 2010.

Bettelheim B. (1967). *The Empty Fortress: Infantile Autism and the Birth of the Self*. New York, NY: The Free Press.

Buxbaum JD. (2009). Multiple rare variants in the etiology of autism spectrum disorders. *Dialogues Clin Neurosci*. 11(1):35–43.

Chew K. (2009). Eugenics, fear, and pain. change.org Web site. http://healthcare.change.org/blog/view/eugenics_fear_and_pain. Updated May 18, 2009. Accessed June 25, 2010.

Coukell A. (2006). You can hurry science. *Proto*. Winter;26–31.

Crenson M. (2008). Anne and Linda unveil 23andWe at D6. The Spittoon [blog]. http://spittoon.23andme.com/2008/05/29/anne-and-linda-unveil-23andwe-at-d6/. Updated May 29, 2008. Accessed June 25, 2010.

Dawson Interview. Interview with Geraldine Dawson. October 10, 2007, Seattle, WA.

Durand CM, Betancur C, Boeckers TM, et al. (2007). Mutations in the gene encoding the synaptic scaffolding protein SHANK3 are associated with autism spectrum disorders." *Nat Genet*. 39(1):25–27.

Folstein S, Rutter M. (1977). Genetic influences and infantile autism. *Nature*. 265: 726–728.

Geschwind Interview. Interview with Daniel Geschwind. May 20, 2009. Los Angeles, CA.

Goldenberg AJ, Hull SC, Botkin JR, Wilfond BS. (2009). Pediatric biobanks: approaching informed consent for continuing research after children grow up. *J Pediatr*. 155(4):578–583.

Greenfeld KT. (2010). The autism debate: who's afraid of Jenny McCarthy? *TIME* Web site. http://www.time.com/time/nation/article/0,8599,1967796-1,00.html. Updated February 25, 2010. Accessed June 25, 2010.

Hull SC, Sharp RR, Botkin JR, et al. (2008). Patients' views on identifiability of samples and informed consent for genetic research. *Am J Bioeth*. 8(10):62–70.

Iversen P. (2006). *Strange Son: Two Mothers, Two Sons, and the Quest to Unlock the Hidden World of Autism*. New York, NY: Penguin Group.

Iversen Interview. Interview with Portia Iverson. April 17, 2009. Los Angeles, CA.

[JDRF] Juvenile Diabetes Research Foundation International. (2010). Juvenile Diabetes Research Foundation International Web site. http://www.jdrf.org/. Accessed June 25, 2010.

Kalb C. (2009). Erasing autism. *Newsweek* Web site. http://www.newsweek.com/2009/05/15/erasing-autism.html. Updated May 16, 2009. Accessed June 25, 2010.

Kaufman DJ, Murphy-Bollinger J, Scott J, Hudson KL. (2009). Public opinion about the importance of privacy in biobank research. *Am J Hum Genet*. 85(5):643–654.

Kaye J, Heeney C, Hawkins N, de Vries J, Boddington P. (2009). Data sharing in genomics—re-shaping scientific practice. *Nat Rev Genet*. 10(5):331–335.

Kirby D. (2009). Top federal autism panel votes for millions in vaccine research. The Huffington Post Web site. http://www.huffingtonpost.com/david-kirby/top-federal-autism-panel_b_155293.html. Updated January 5, 2009. Accessed June 25, 2010.

Kohane IS, Mandl KD, Taylor PL, Holm IA, Nigrin DJ, Kunkel LM. (2007). Medicine. Reestablishing the researcher-patient compact. *Science*. 316(5826):836–837.

LaJonchere Interview. Interview with Clara LaJonchere. October 10, 2008. Phone interview.

LaJonchere Communication. Oral Communication. July 16, 2010.

Mailman MD, Feolo M, Jin Y, et al. (2007). The NCBI dbGaP database of genotypes and phenotypes. *Nat Genet*. 39(10):1181–1186.

McGuire AL, Hamilton JA, Lunstroth R, McCullough LB, Goldman A. (2008). DNA data sharing: research participants' perspectives. *Genet Med*. 10(1):46–53.

Muhle R, Trentacoste SV, Rapin I. (2004). The genetics of autism. *Pediatrics*. 113(5): e472–e486.

Neergaard L. (2010, February 8). Ethics debate over blood from newborn safety tests. *The Boston Globe*. http://www.boston.com/news/health/articles/2010/02/08/ethics_debate_over_blood_from_newborn_safety_tests/. Accessed July 6, 2010.

Orsini M. (2009). Contesting the autistic subject: biological citizenship and the autism/autistic movement. In: Murray SJ, Holmes D, eds. *Critical Interventions in the Ethics of Healthcare: Challenging the Principle Authority in Bioethics*. Surrey, UK and Burlington, VT: Ashgate Publishing: pp. 115–130.

Pinto D, Pagnamenta AT, Klei L, et al. (2010). Functional impact of global rare copy number variation in autism spectrum disorders. *Nature*. 466(7304):368–372.

Regalado A. (2005, December 15). A hedge-fund titan stirs up research into autism. Derived from Regalado A. Titan's millions stir up research into autism: James Simons taps big stars from outside field to find a genetic explanation. *Wall Street Journal*. post-gazette.com Web site. http://www.post-gazette.com/pg/05349/622925.stm. Updated December 15, 2005. Accessed November 19, 2010.

Risch N, Spiker D, Lotspeich L, et al. (1999). A genomic screen of autism: evidence for a multilocus etiology. *Am J Hum Genet*. 65(2):493–507.

Sharp RR, Yarborough M, Walsh JW; Ethical, Legal, Social Issues Working Group of the Alpha-1 Foundation. (2008). Responsible patient advocacy: perspectives from the Alpha-1 Foundation. *Am J Med Genet A*. 146A(22):2845–2850.

Shestack Interview. Interview with Jon Shestack. April 17, 2009. Los Angeles, CA.

Silverman C. (2004). *A Disorder of Affect: Love, Tragedy, Biomedicine, and Citizenship in American Autism Research, 1943–2003* [dissertation]. Pittsburgh: University of Pennsylvania.

Solovitch S. (2000, June 18). The stolen child. *SV Magazine*. Sara Solovitch Web site. http://www.sarasolo.com/sv.html. Accessed November 19, 2010.

Szatmari P, Jones M. (2007). Genetic epidemiology of autism spectrum disorders. In: Volkmar FR, ed. *Autism and Pervasive Developmental Disorders*. 2nd Ed. Cambridge, UK: Cambridge University Press: pp. 157–178.

Taylor G. (2007). Does Autism Speaks support autism eugenics? Adventures in Autism [blog]. http://adventuresinautism.blogspot.com/2007/10/does-autism-speaks-support-autism.html. Updated October 24, 2007. Accessed June 25, 2010.

Terry SF, Terry PF, Rauen KA, Ulitto J, Bercovitch LG. (2007). Advocacy groups as research organizations: the PXE International example. *Nat Rev Genet*. 8(2): 157–164.

Trinidad SB, Fullerton SM, Bares JM, Jarvik GP, Larson EB, Burke W. (2010). Genomic research and wide data sharing: views of prospective participants. *Genet Med*. 12(8):486–495.

Wexler A. (1996). *Mapping Fate: A Memoir of Family, Risk, and Genetic Research*. Berkley: University of California Press.

Winickoff DE, Winickoff RN. (2003). The charitable trust as a model for genomic biobanks. *N Engl J Med*. 349(12):1180–1184.

Commentary on the Discovery Phase of Research

SARA GOERING, SUZANNE HOLLAND, AND KELLY EDWARDS

The discovery phase of the translational pathway traditionally emphasizes basic science research and, in the realm of genomics research, the identification of relevant genetic markers for disease susceptibility, therapeutic response, or potential therapeutic targets. Given the common assumption that basic science is objective and unbiased, this particular phase of the translational pathway, on the face of it, might seem to have little in common with issues of justice as we outline them in Chapter 1. A common claim is that basic scientific knowledge is valuable in itself, and ethical issues arise later, primarily with respect to how we *use* that knowledge. However, norms in the *production* of basic scientific knowledge—standard practices and ways of framing problems—can lend themselves to very particular kinds of understanding (Fox Keller 1995; Longino 1990), and thus lead to gaps in our ability to translate effectively and equitably.

All of the elements of responsive justice (distribution, recognition, and responsibility) are salient in the discovery phase, but in this commentary we focus on issues of what we have called the *recognition* element of justice, defining it as "a reciprocal relation of respectful engagement and attentive concern that allows for shared power and parity of participation" (Goering, Holland, and Fryer-Edwards 2008, p. 46). Recognition has to do with inclusion of a variety of stakeholders in research, but also with guarding against *mis*recognition of those who might readily be overlooked when science is intent on "bench discovery." The concern, in other words, is about not simply inclusion of minority groups, but also responding to the reasons some such groups may feel threatened by such inclusion. Misrecognition, as we have pointed out, often reflects deeper structural issues in traditional research, rather than simply lack of awareness on the part of researchers. For example, when genetic researchers pursued migration theory studies with the Havasupai samples, a "business as usual" practice among genetic scientists, the researchers did not appreciate the impact this investigation would have on the tribe's self-concept (Harmon 2010).

At the discovery phase, therefore, it is both possible and desirable to highlight normative questions that focus on inclusion (what genetic samples get used, how are they obtained, why do certain groups participate or decline to participate, etc.), control over data (how data are housed and shared, who has a say in the control mechanisms, when preliminary data is satisfactory to move toward a handoff to the development phase), and allocation of scarce resources (what are the most productive areas of study (and productive in what ways), who will most likely benefit from the advances). Attending to recognition requires research strategies that reflect the actual needs and interests of those the research is intended to benefit—and, by extension, a specific obligation to seek and address the concerns of groups traditionally marginalized or underserved.

In the discovery-phase chapters in this book, the authors show how basic factors in the organization of research (whose samples get included in discovery research; who controls data, and when can it be shared between researchers) can significantly affect the possibility of successful, equitable translation. Both chapters raise questions about the distribution of the benefits of genetic translational research, given standard practices in basic science, and signal the significance of researcher responsibility for reassessing and possibly transforming engrained scientific practices. Chapter 3, for example, demonstrates several ways in which members of ethnic minority groups may not benefit as readily from genetic testing as their European American counterparts, given their undersampling in basic research. It also highlights the complexity in figuring out a remedy for that problem. Chapter 4 shows how community advocacy, even at this early stage of the pathway, can transform standard research practices in ways that may advance translation and, at the same time, empower individuals in communities affected by the health burden, even in the absence of a cure or reliable genetic test.

Fullerton (Chapter 3) argues that because the bulk of genetic samples used in basic genetic research come from people of European American ancestry, individuals from ethnic minority groups may find that genetic tests offered in the marketplace, or in a clinician's office, are much less helpful for them due to the increased likelihood of false negatives, false positives, or simply inadequate information. As she suggests through a series of hypothetical scenarios, a lack of diversity in the samples put into the front end of the translational pathway affects the power of the downstream tests to help traditionally underserved populations. This "input-output" problem points to the possibility of distributional inequities in respect to the benefits of genetic testing, not simply because of who can afford the tests or who has access to health coverage that will include them, but also because of what the tests can actually tell individuals.

One response to this concern about fair distribution downstream might be to advocate greater inclusion of genetic samples from ethnic minority populations in the basic science. As Fullerton notes, however, even if community distrust of genetic investigations could be overcome, proportional inclusion of samples from all population ancestry groups would not alter the inadequacy of the tests; *over*-sampling of ethnic minority groups would be required to provide the requisite power to determine disease markers within ethnic minority groups. Furthermore, the efficacy of tests for individuals who are unsure of their genetic ancestry groups may still be questionable.

While we advocate that researchers keep in mind the importance of the question of "inclusion," Fullerton deepens the analysis by promoting benefits-based sampling over simple inclusion, which is to say that researchers need to recognize how the size of relevant ancestry groups may affect the results of their research and then aim to include proportionally more samples from underserved ethnic minority groups so as to create tests that are meaningful for them. A related concern arises when research is extended to include environmental contributors to disease and gene–environment interactions. Environmental exposures differ across social groups. Research outcomes can be influenced not only by the size of the sample, but also by how well it is characterized, including how fully environmental exposures are measured.

Though genomics researchers may not themselves be well placed to find, partner with, and negotiate the possibility of sample provisions from minority or other underserved groups, they could look to cross-disciplinary collaborations with academic colleagues who have existing relationships with relevant community groups. Recognizing the importance of expanding their sample base, and thus altering "business as usual" in the lab, is, from a justice perspective, a responsibility researchers should take on. Though current practices further scientific understanding and create possibilities for translation to improved health outcomes, the needs of underserved populations are likely to continue to go unaddressed on this model unless research practices begin to reflect more of the recognition element of justice.

Scientists have a responsibility to address issues of recognition, but they may not yet have the relevant incentive structures to take up this challenge. Such incentives can be developed in part through advocacy by community stakeholders. An example of how incentive structures for the transformation of research practices might be encouraged is offered by Tabor and Lappé in their chapter on how advocacy by parents of children with autism changed data-sharing practices in genetic autism research (Chapter 4). Of course, the parent advocates described in relation to the autism debate are relatively powerful in respect to social and political capital; tribal groups like the Havasupai, who have been politically disenfranchised for hundreds of years, are now asserting their claims through strong advocacy, but they and other marginalized groups might face more imposing barriers to garnering the social capital needed to transform standard research practices. This reality highlights the necessity of researchers' taking responsibility for ensuring recognition, whether or not they intend to partner with underserved populations.

Tabor and Lappé tell the story of Autism Genetic Resource Exchange (AGRE) and its ability to influence standard research practices once the perspectives of parents—who felt the urgency for a better understanding of autism, given their personal connections with children with the disease—were fully recognized and allowed to share some control over the discovery research done by scientists in the field. As Tabor and Lappé point out, this is a success story with respect to data sharing and recognition of nonscientific community interests, but these changes have not in fact led to the end desired by the parent advocates, namely, a cure or biomedical therapy for autism. Improvements in the translational pathway will not, of necessity, result in translational success stories. Indeed, with respect to the struggle for justice in this arena, recognition might call for more attention to the voices of other relevant stakeholders in the autism debates. For instance, autistic individuals who criticize the current emphasis on genetic research and curative therapies—members of the neurodiversity movement—have demanded respect for their ways of being in the world and argued for more research and/or services focused on improving the well-being of existing people across the spectrum of autism, rather than insistence on a cure (Ne'eman 2010). The recently formed Academic Autistic Spectrum Partnership in Research and Education (AASPIRE 2010) urges a community-based participatory research model for work on autism, to ensure that academics and clinicians who are interested in making their careers in relation to autism have a solid grasp of what autistic individuals themselves have to say about their condition and what might benefit them.

Through these two chapters, we see the importance of partnerships or, at a minimum, scientists coming to apprehend the needs and interests of the target research community. Here at the beginning of the translational cycle, getting these questions of justice and benefit right helps guide the research in the best possible direction.

REFERENCES

[AASPIRE] Academic Autistic Spectrum Partnership in Research and Education. (2010). Community based participatory research. AASPIRE Web site. www.aaspireproject.org/about/cbpr.html. Accessed June 14, 2010.

Fox Keller E. (1995). *Refiguring Life: Metaphors of Twentieth Century Biology*. New York, NY: Columbia University Press.

Goering S., S. Holland, K. Fryer-Edwards (2008). Transforming Genetic Research Practices with Marginalized Communities: A Case for Responsive Justice. *Hastings Center Report* 38 (2) 43–53.

Harmon A. (2010, April). Where did you go with my DNA? *New York Times*. http://www.nytimes.com/2010/04/25/weekinreview/25harmon.html Accessed January 19, 2011

Longino H. (1990). *Science as Social Knowledge: Values and Objectivity in Scientific Inquiry*. Princeton, NJ: Princeton University Press.

Ne'eman A. (2010). The future (and the past) of autism advocacy, or why the ASA's magazine, *The Advocate*, wouldn't publish this piece. *Disability Studies Quarterly*. 30(1). www.dsq-sds.org/article/view/1059/1244, Accessed May 10, 2010.

5

Early Assessment of Translational Opportunities

PATRICIA DEVERKA AND DAVID L. VEENSTRA

The translation of novel scientific discoveries into clinically useful applications represents one of the most significant hurdles in medical science. In the United States, federal agencies including the National Institutes of Health (NIH) (Zerhouni 2007) and the Food and Drug Administration (FDA 2004) have sought to address these challenges with infrastructure and project funding designed to speed the clinical integration of new discoveries that deliver health benefit to patients. The regulatory determination of which candidate health applications move into clinical practice typically depends on whether there are positive results from well-designed clinical studies with clinically relevant endpoints. The clinical studies that help ensure patient safety and establish efficacy—particularly for innovations that are more invasive or pose higher or unknown risks to patients—are costly. Private-sector companies, which stand to generate a positive return for investors if such products gain regulatory approval and are successfully marketed, are often the major sponsor of these clinical studies. Private companies also fund much discovery research but, unlike public institutions, typically take the lead in conducting medical product development studies and navigating the regulatory process (PhRMA 2005).

However, regulatory agencies are not the only arbiters of whether a new medical product will be used in clinical practice; their threshold determination should be viewed as a necessary, but not sufficient, condition for clinical integration. Clinical experts and their associated professional societies, as well as experts in evidence-based medicine, typically conduct systematic reviews of the published literature supporting specific clinical applications for the technology, with the intention of making evidence-based practice recommendations (Guyatt et al. 2000).

These assessments can be critically important for helping to guide clinical practice, as well as for informing third-party payer decision-making. In the United States, public or private insurance coverage of a new medical technology is typically required before an individual can realistically gain access to the new product or service, as most people are unable (or unwilling) to pay for medical care out of pocket. Payers often rely on evidence-based technology assessments when evaluating whether the clinical value of a new technology has been established and therefore merits coverage or, alternatively, if the current evidence base is lacking and the intervention should be excluded from coverage (Garber 2001).

Therefore, for most promising new medical technologies that enter the development phase of translation, there are *three major target audiences* for scientific evidence that typically influence the research and investment portfolio supporting their clinical integration. The first audience includes the various regulatory agencies, which emphasize demonstration that the benefits outweigh the harms of the new technology. The second audience consists primarily of clinicians, scientific experts, and professional groups who tend to focus on determining whether there is sufficient evidence to suggest that use of the new technology improves health outcomes for patients as compared to current practice. The third group consists of payers who attempt to use their purchasing power to promote value for money (cost-effectiveness) in medical care by demanding robust scientific evidence that the presumed effectiveness of the new intervention has been adequately demonstrated for one or more target patient groups. Payer scrutiny is particularly stringent for interventions that target prevalent conditions, substantially change current practice patterns, or have a high unit cost.

For genetic tests, the development phase of translational research begins with the identification of a candidate health application, continues through the time period when the genetic test becomes available for clinical use and when coverage and reimbursement decisions are being made, and ends when sufficient evidence exists to develop clinical practice guidelines (Khoury et al. 2007). The development phase can be quite long and complicated, even for an innovation such as pharmacogenetic (PGx) testing, which is often described as one of the most promising and near-term applications of the Human Genome Project (Collins and McKusick 2001; Evans and Relling 1999; Swen et al. 2007). Despite an abundance of published studies and marketed genetic tests, a number of disease-risk and PGx tests that have undergone systematic evidence reviews have encountered substantial barriers to clinical integration (see Table 5-1).

Delays and disconnects in the development phase can have important effects on pharmaceutical manufacturers and clinicians, but also—and most importantly—on patients. Where tests with proven effectiveness are commercially available but are not used by clinicians, patients potentially miss out on important benefits and may be harmed by suboptimal treatment. For example, a PGx test can help clinicians determine whether a woman's breast cancer will respond to Drug X. If she has the test, optimal therapy can be instituted right away, whereas if she does not have the test she may receive several rounds of less-effective chemotherapy—along with its side effects. However, premature translation of genetic tests may also be harmful

Table 5-1. Genetic Tests that Have Undergone Evidence Review by EGAPP Working Group (EWG)

Test	Review findings	Recommendation
CYP2D6 testing to guide treatment with SSRI antidepressants	*Studies did not consistently identify a significant association between CYP450 genotype and clinical response to SSRI treatment or adverse events*; studies were generally small and of poor quality	Insufficient evidence to support a recommendation for or against use of test
UGT1A1 testing to guide irinotecan treatment in colon cancer patients with the intent of modifying the dose to avoid adverse drug reactions (severe neutropenia)	• *Adequate evidence* of a significant association between *UGT1A1* genotype and the incidence of severe neutropenia at standard doses of irinotecan • *No evidence* to support clinical utility in the proposed clinical scenario • Preliminary modeling suggests that, even if targeted dosing were to be highly effective, it is not clear that benefits (reduced adverse drug events) outweigh harms (unresponsive tumors)	Evidence is currently insufficient to recommend for or against the routine use of test

Gene expression profiling to guide adjuvant chemotherapy treatment decisions in early-stage breast cancer	• Overall, evidence of analytic validity was heterogeneous • *Adequate evidence* regarding the association of the Oncotype DX Recurrence Score with disease recurrence • *Adequate evidence* for response to chemotherapy • *Adequate evidence* to characterize the association of MammaPrint with future metastases, but *inadequate evidence* to assess the added value to standard risk stratification; could not determine the population to which the test would best apply	• Insufficient evidence to make a recommendation for or against the use of tumor gene expression profiles • Preliminary evidence of potential benefit of Oncotype DX for some women, but could not rule out the potential for harm to others
Genetic testing to identify patients with Lynch syndrome at higher risk of developing colon cancer	• *Adequate evidence* to describe the clinical sensitivity and specificity for three preliminary tests, and for four selected testing strategies • *Limited but promising evidence* suggesting that testing can improve outcomes	Sufficient evidence to recommend offering genetic testing for Lynch syndrome to individuals with newly diagnosed colorectal cancer to reduce morbidity and mortality in relatives

to patients. When tests are marketed based primarily on association or retrospective data, the lack of evidence supporting PGx-guided treatment decisions is often not clear to patients and their providers. The situation is particularly problematic in the setting where PGx testing is marketed directly to consumers; in these circumstances, patients may feel anxiety about whether to request a test, share test results with their doctor, or follow the course of action suggested by the test. Potentially, they could even make decisions about their drug therapy that could result in a decrease in clinical benefit or an increase in the risk of an adverse event. Thus, addressing the challenges of the development of genetic tests will be important both at the individual patient level and from a population-wide public health perspective.

At this stage of translation, the root of the problem is the lack of incentives to develop an adequate evidence base prior to a genetic test's becoming available for clinical use. Much of the work is conducted by the private sector and is focused on crossing existing regulatory hurdles for test approval, which many stakeholders believe have been set too low. Decisions at this stage of translation are often (but not exclusively) dictated by the commercial interests of both large and small test developers. In the absence of formal regulatory requirements to support the demonstration of clinical utility, most genetic tests diffuse into the marketplace with poorly characterized risk/benefit profiles and a great deal of uncertainty regarding how the tests should be used in clinical practice. There also are insufficient marketplace returns to spur investment in clinical utility studies, given that diagnostic tests have historically been reimbursed on a cost basis rather than a value basis. The absence of financial rewards on par with the costs of the research investment makes it unlikely that developers alone will solve the problem of lack of evidence.

The compelling challenge at this stage of translation is how to balance the well-intentioned desire for better evidence with the potential for creating significant barriers to innovation. If the evidence bar is set too high, there may be fewer incentives for companies and venture capitalists to invest in genetic test research and development. If the evidence bar is set too low, patients may be harmed by the use of inadequately studied tests, and health care resources will be wasted on ineffective interventions. At the time of product launch, the cost-effectiveness of a new genetic test is typically unknown, and payers become the ultimate arbiters of whether a test will be available for clinical use through their coverage and reimbursement decision-making processes.

Given the large number of studies that are frequently found when a literature search is conducted about a particular genetic test, the issue is not necessarily that more studies need to be done; however, it is clear that *different* studies need to be done if we are to understand how use of the test impacts clinical outcomes (Tatsioni et al. 2005). Comprehensive guidance principles for designing studies that minimize the sources of bias commonly found in the literature have been described (Pepe et al. 2008). However, today it is atypical for most marketed genetic tests to be supported by a peer-reviewed evidence base that addresses the questions of greatest relevance to clinicians, patients, and payers.

We argue that this phase of translation—development—represents a leverage point in the evidence chain. Additional studies can be conducted both premarketing and postmarketing that will likely pay significant dividends in terms of advancing our understanding of the clinical utility (net balance of risks and benefits of using the test in clinical practice) of genetic testing. Aligning the incentives at this phase of translation is and should be a major focus of efforts to enhance the appropriate diffusion of new genetic tests into clinical practice.

CASE STUDY: PHARMACOGENETIC TESTING FOR WARFARIN RESPONSE

The use of PGx testing to guide induction of warfarin therapy is an illustrative example of the difficulties that can hinder the integration of new genomic innovations into clinical practice. Numerous studies on this topic have been published, and warfarin PGx testing is clinically available, but substantial controversy remains about whether or how testing should be used in clinical practice. As a result, very few payers have agreed to reimburse the cost of testing. Despite a great deal of enthusiasm in the research community for the association between genetic variation and interindividual responsiveness to warfarin—and despite the large, unmet medical need to improve the safety profile of warfarin therapy—there has been very little uptake of PGx testing to guide warfarin use in clinical practice. In this section, we will use the example of warfarin PGx testing to illustrate the role of federal regulatory requirements regarding analytic validity, clinicians' and specialty societies' demand for evidence of clinical utility, and payers' need for cost-effectiveness data in whether, or how, genomic innovations achieve clinical integration in the United States.

Warfarin and Pharmacogenetic Testing

Warfarin is an oral blood-thinning drug commonly used to prevent thromboembolic events (clots) in patients with a history of previous clots, health conditions such as atrial fibrillation (heart flutter), and artificial heart valves. Warfarin has a "narrow therapeutic index"—too high a dose can lead to major bleeding, and too low a dose does not protect from clotting events. In addition, there is high variability in response, both between patients and within an individual patient over time. Warfarin therapy is therefore carefully managed using a blood test called the International Normalized Ratio (INR), which is used to monitor anticoagulation response. Typically patients undergo this blood test and corresponding warfarin dose adjustments every one to three days initially, and then every two to six weeks. The starting dose for warfarin is based on patient characteristics (age, sex, co-medications, liver function, etc.) and is often about 5 mg/day.

Variants of two genes have been shown to have a significant effect on warfarin dose requirement: cytochrome P450 2C9 (*CYP2C9*, encoding for the principal enzyme in warfarin metabolism) and vitamin K epoxide reductase complex 1 (*VKORC1*, encoding for the target protein inhibited by warfarin) (Limdi and Veenstra 2008). A warfarin-dose prediction algorithm using clinical, demographic, and genomic information was recently developed by the International Warfarin Pharmacogenetics Consortium (IWPC); it incorporated data from 5,700 patients from nine countries (IWPC et al. 2009). Dose prediction including pharmacogenetic information improved the ability to accurately predict patients requiring ≤ 3 mg/day (54.3% versus 33.4%) and those requiring ≥ 7 mg/day (26.4% versus 9.1%), compared to prediction using clinical and demographic information only. However, the impact of genotype-guided dosing on clinical outcomes such as INR values and bleeding and clotting events is less clear. The two small clinical studies conducted to date have not been conclusive (Anderson et al. 2007; Caraco, Blotnick, and Muszkat 2008).

Thus, the use of genomic information *may* improve clinicians' ability to predict the patient's response to warfarin, thereby minimizing the time it takes to find a stable dose and potentially decreasing the risk of bleeds or clots—but to date there is no solid clinical evidence to confirm this (Eckman et al. 2009; Schwarz et al. 2008). A recent set of commentaries by Larry Lesko, Director of the Office of Pharmacology at the FDA, and David Garcia, Department of Internal Medicine at the University of New Mexico and President of the AntiCoag Forum clinician group, highlights the contrasting view of some of the stakeholders. Lesko states:

> The question about warfarin pharmacogenetics before us now is not "is it ready for prime time?" The more important question is, while more and more studies are being planned and/or conducted, should we accept and use our current knowledge about genetic factors to improve the quality of warfarin initial dosing and anticoagulation in our patients. The benefits and risks of pharmacogenetics, in my view, favor pharmacogenetics. (Lesko 2008, p. 303)

He ends by quoting Voltaire: "The perfect is the enemy of the good." (p. 303)

Garcia, in contrast, states:

> There is the potential that this testing could result in net harm to patients. For example, a result of "wild type" from the genetics laboratory may impart an unjustified sense of security to the clinician . . . I remain highly uncertain about whether (or in what settings) the addition of genetic testing to current practice will reduce the risk of hemorrhage and/or thrombosis among warfarin-treated patients . . . the most important final conclusions must come from randomized, controlled comparisons. (Garcia 2008, p. 305)

The solid evidence of clinical validity, unclear evidence of clinical utility, and, in particular, contrasting perspectives of stakeholders of the actual benefit of warfarin pharmacogenomic testing make it an ideal case study for this chapter. Furthermore, as can be seen in Figure 5-1, the timeline for the development stage

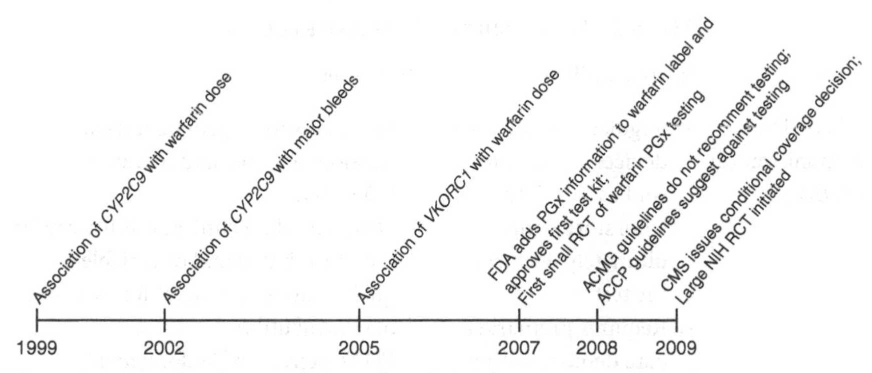

Figure 5-1 Timeline for translation of warfarin pharmacogenetic testing.

for warfarin pharmacogenetics spans 10 years. The lessons learned from examination of this case study will likely be informative for a range of genetic tests that have faced similar roadblocks when introduced to the marketplace.

REGULATORY REQUIREMENTS

There are two possible regulatory pathways for obtaining FDA approval for a genetic test, depending on whether it is developed as a test kit or as a laboratory service (SACGHS 2008). A minority of genetic tests are developed as test kits, which are viewed as medical devices and are required to undergo premarket review by the FDA under its authority to regulate in vitro diagnostic devices. More commonly, genetic tests are developed as laboratory services and come to market as laboratory-developed tests. In the latter situation, it is the responsibility of the Centers for Medicare & Medicaid Services (CMS) to implement and enforce the Clinical Laboratory Improvement Amendments of 1988 (CLIA), the statutes designed to ensure the quality of testing performed at clinical labs. All other things being equal, the regulatory hurdles are typically lower for laboratory-developed tests than for genetic tests that undergo premarket review by the FDA (see Table 5-2).

There are a number of PGx tests for warfarin available on the market today, including both FDA-approved test kits (five as of this writing) and laboratory-developed tests offered by individual clinical laboratories (CMS 2009). The tests measure both the wild-type genotype and the single-nucleotide polymorphisms that are related to differential sensitivity to warfarin. However, as there are no practice or regulatory guidelines to standardize the subset of particular polymorphisms that must be measured or the measurement approach, manufacturers are free to decide which polymorphisms for *CYP2C9* and *VKORC1* are most clinically relevant (Rosove and Grody 2009). This lack of consistency in measurement from test to test complicates the comparison of various assays and the evaluation of clinical algorithms that incorporate different warfarin-specific tests.

Table 5-2. HOW GENETIC TESTS ARE REGULATED

Agency	Responsibilities	Problems
Food and Drug Administration (FDA)	• Regulates (as medical devices) genetic tests sold as "kits"[1] to laboratories that ultimately perform the test • Requires premarket data demonstrating both analytic and clinical validity for genetic test kits • Has regulatory authority to oversee laboratory-developed tests	• Vast majority of genetic tests are not developed as kits and therefore avoid FDA review • Data regarding clinical validity can be literature-based and of variable quality; no requirement for evidence of clinical utility • FDA's exercise of "enforcement discretion" has historically meant the agency does not involve itself in the oversight of these tests developed in-house by clinical laboratories
Centers for Medicare and Medicaid Services (CMS)	• Certifies analytic validity of genetic tests performed by clinical labs under Clinical Laboratory Improvement Amendments of 1988 (CLIA) • Requires proficiency testing for "high-complexity tests" when classified as a specialty area under CLIA	• CLIA does not authorize CMS to evaluate clinical validity of a genetic test or to require pre- or postmarket approval; decision to offer test is made solely by the laboratory director • Although genetic tests are considered "high complexity," because CMS has not followed through on recommendations to create a genetic testing specialty area, there are no standard proficiency testing programs for genetic tests

In 2007, the FDA updated the drug label for warfarin to include information about the association between *CYP2C9* and *VKORC1* variation and warfarin responsiveness. This change was based on an assessment of published studies that established that approximately 30% of the variability in responsiveness to warfarin can be traced to variants in these two genes, combined with the FDA's enthusiasm for the potentially large public health benefit of improving warfarin's safety profile (Kim et al. 2009). The label change simply informs the prescriber about the association between genotype and warfarin-dosing requirements. It does not require pharmacogenetic testing prior to initiating warfarin therapy.

1. Test kits contain the reagents needed to perform the tests, instructions on test performance, and information regarding what mutations are detected.

CLINICAL UTILITY

The regulations governing PGx test development primarily emphasize the demonstration of analytic validity. Less attention is given to the issue of clinical validity, that is, how accurately the test detects or predicts a specific clinical disorder or outcome. The regulations thus enable rapid market entry while tending to disincentivize investment in more rigorous studies of clinical validity. And evidence of clinical utility (i.e., whether the test improves clinical outcomes and whether testing provides a meaningful incremental benefit in patient-management decision-making) is not a regulatory requirement. It is uncommon to find published evidence of how use of a given genetic test improves clinical care. Studies designed to demonstrate clinical utility are expensive to conduct, and most diagnostic companies are unable to finance such studies without major changes to their current business models. Yet this is exactly the type of evidence that is needed by multidisciplinary expert panels that review the evidence base for a wide range of genetic tests, such as the Evaluation of Genomic Applications in Practice and Prevention (EGAPP) Working Group (Teutsch et al. 2009), as well as large, influential payers such as the CMS (CMS 2009) and Blue Cross and Blue Shield Association companies (BCBS 2009).

Different stakeholders in the test-development arena can have dramatically different perceptions of potential clinical benefit. Expert clinicians generally believe that available evidence is insufficient to warrant routine use of warfarin PGx testing, and that additional research is needed. In 2006, the American College of Medical Genetics issued a policy statement based on an evidence-based review for warfarin PGx (Flockhart et al. 2008). The policy states that "there is insufficient evidence, at this time, to recommend for or against routine CYP2C9 and VKORC1 testing in warfarin-naive patients." (p. 139). Two years later, the American College of Chest Physicians, which develops the definitive evidence-based guidelines for anticoagulation treatment, stated, "[W]e suggest *against* pharmacogenetic-based dosing until randomized data indicate that it is beneficial (Grade 2C)" (Ansell et al. 2008, p. 161S). Most recently, a systematic review by Kangelaris and colleagues (2009) also reported lack of evidence of benefit.

The perspective from stakeholders offering testing services (e.g., test manufacturers and testing providers) is somewhat different. Howard Coleman, CEO of Genelex, a testing provider, stated:

> In summary, the weight and quality of evidence supports genetic testing for *2C9* and *VKORC1* gene variants in elderly patients before or shortly after starting warfarin to improve the quality of anticoagulation and define starting doses that are closer to the eventual stable maintenance doses of warfarin. (DNADiva 2008)

And as noted above, Larry Lesko of the FDA has argued that randomized, controlled trials (RCTs) "proving" benefit are not always needed when there is sufficient plausibility of benefit.

In summary, different stakeholders endorse different evidence thresholds, which in turn affect the motivation for the generation of additional evidence. Clinicians, test developers, and regulatory authorities appear to be at loggerheads. To address this issue, the NIH has invested in the Clarification of Optimal Anticoagulation Through Genetics (COAG) study, an RCT comparing the use of warfarin pharmacogenomic testing to use of a clinical algorithm to guide warfarin dosing. The trial will enroll approximately 1,200 patients. It began enrollment in 2009, with results scheduled for 2011–2012 (Clinicaltrials.gov 2009).

REIMBURSEMENT

Another component of the assessment of net benefits and harms of a new genetic test is the evaluation of its economic value. PGx testing promises to make prescribing more efficient in the immediate term (through the avoidance of potentially unsafe or ineffective medications) as well in the longer term (because patients who receive targeted drug treatments are more likely to experience improved health outcomes). However, payers are concerned that these tests will actually increase health care costs because they are poorly studied and their utilization is difficult to track (Davis et al. 2009).

One of the most significant barriers to generating cost-effectiveness data for PGx tests is the antiquated U.S. reimbursement system for laboratory tests, which focuses on the technical comparability of new versus old tests. Under the resulting classification scheme, payers typically set reimbursement rates on a cost basis rather than a value basis, which offers few market-based incentives for developers to generate evidence that use of the genetic test improves outcomes for patients. (Higher reimbursement rates would potentially provide the return on investment that would justify investment in expensive outcomes studies.) Without comparative effectiveness data, there is no objective way to measure the presumed benefits of a more expensive genetic-testing strategy that may ultimately lead to better outcomes for patients and the health care system.

There have been several formal analyses of the cost-effectiveness of warfarin pharmacogenomic testing. As noted above, there are no direct data of the benefit of warfarin pharmacogenomic testing, so these analyses inherently involve assumptions regarding the benefits and harms. The most notable report published to date is a cost-effectiveness study conducted by health economists at the FDA, which was publicly released in 2006 via the Web site of the American Enterprise Institute-Brookings Joint Center for Regulatory Studies (McWilliam, Lutter, and Nardinelli 2006). For modeling purposes, the authors assumed that testing would completely eliminate the risk of excess bleeding resulting from genetic variation (i.e., 100% effectiveness) and reduce clotting risk significantly as a result of improved anticoagulation control. The authors stated, "We estimate the reduced health care spending from integrating genetic testing into warfarin therapy to be $1.1 billion annually, with a range of about $100 million to $2 billion" (McWilliam,

Lutter, and Nardinelli 2006, p. iii). This study, which was not well received by the clinical community, has been criticized for its optimistic assumptions (Hughes and Pirmohamed 2007; Veenstra 2007). Clinicians' negative response may also have been influenced by the test providers' decision to use this study as part of their promotional materials.

A number of cost-effectiveness studies that have come to more conservative conclusions about the value of warfarin PGx testing have recently been published in the peer-reviewed literature. Eckman and colleagues (2009) used a similar methodological approach to the FDA study; however, they estimated the effect of testing on bleeding risk by pooling data from three small, nondefinitive, clinical trials. The authors concluded that "[w]arfarin-related genotyping is unlikely to be cost-effective for typical patients." (p. 73) Patrick and colleagues developed a cost-effectiveness model based on assumed impacts of testing on anticoagulation control (INR levels) (Patrick, Avorn, Choudry 2009) and suggested, "Given the current uncertainty surrounding genotyping efficacy, caution should be taken in advocating the widespread adoption of this strategy." (p. 429)

Thus, the economic value of warfarin pharmacogenomic testing is uncertain, and the majority of recent studies suggest that it is not cost-effective. However, the cost of testing is rapidly decreasing, which could shift the balance of future analyses. For example, if testing could be obtained for $50 rather than $250, the level of benefit would not need to be large to provide reasonable economic value.

CMS recently issued a coverage decision for warfarin PGx testing that will provide some incentive for evidence generation. The official memo states that while CMS does not believe the available evidence supports coverage of warfarin PGx for Medicare beneficiaries, there is sufficient evidence to support "coverage with evidence development," a classification that pays for testing conducted in the research setting (CMS 2009). The decision specifies that warfarin PGx testing will be reimbursed only for patients initiating warfarin who are enrolled in an RCT that measures major bleeding and thromboembolic events. The Medicare reimbursement for warfarin PGx testing is approximately $270, which is likely insufficient to cover the per-patient cost of conducting such a clinical trial, and other funding sources—both private and public—will be required. The results from multiple moderately sized trials could be pooled, along with the results from the NIH-sponsored COAG trial, to provide a sufficient evidence base for determining the clinical utility and cost-effectiveness of warfarin PGx testing.

Our examination of the circumstances surrounding warfarin PGx testing illustrates several important points. First, new genetic tests are often introduced into clinical practice before evidence that meets the needs of clinicians and payers is available. Second, the current regulatory regime allows test developers to choose whether to bring a new test to market as a test kit (through the FDA) or as a testing service, bypassing the FDA and using the laboratory-developed testing route. At a minimum, this two-track system complicates quality assurance efforts; it also tends to encourage smaller companies and those with limited resources to opt for the laboratory-developed test route to reduce their time to market. Given this state of affairs, it is not surprising that a review of the PGx literature most often reveals studies that support the analytic and clinical validity, rather than clinical utility, of a given test; companies act rationally in their self-interest to meet (but not exceed) the current requirements for gaining approval to market a new test. However, academic researchers also design studies that fall short of demonstrating the impact of PGx testing on patient relevant outcomes, as our case analysis of warfarin PGx testing has demonstrated—although study designs and feasibility are highly correlated with funding opportunities. The need for better understanding of the information needs of payers and clinicians, and the recognition of a pattern of limitations of the publicly available evidence at this critical stage of translation, are the lessons learned from the first wave of genetic tests that have been subject to systematic evidence reviews (see Table 5-1).

The next two factors influencing whether or how genetic tests are integrated into clinical practice—evidence of clinical utility and evidence of economic value—are inextricably linked, as economic value cannot be established without evidence of clinical utility. Our case analysis for warfarin demonstrates that there is often a range of opinion regarding whether the current evidence base supports the use and reimbursement of PGx testing. However, the majority of professional groups, expert review panels, and payers have concluded that PGx testing for warfarin should *not* be routinely recommended at this time, precisely because there is a lack of data to demonstrate that testing has a positive impact on patient-relevant outcomes. Indeed, the lack of evidence of clinical utility has been identified as one of the major barriers to clinical translation by every major federal group that has reviewed the field of genetic testing (SACGHS 2009). The history of warfarin PGx testing illustrates that the regulatory framework for approval of a PGx test is insufficient to ensure clinical adoption, and that evidence of clinical utility will be required before there is professional and payer consensus about PGx-guided changes in areas of established clinical practice. Certainly, this evidence threshold will vary depending on the particular disease state and the relative contribution of PGx testing to predicting treatment response, but it seems reasonable to expect that the pathway for clinical translation will only be optimized if studies are designed to address this information need.

POLICY IMPLICATIONS

The three primary drivers discussed in this chapter that can influence evidence generation at the development stage of the translational pathway—regulatory policy, clinical utility, and reimbursement—face significant impediments: (1) unclear or lacking regulatory policy, (2) varied perceptions of clinical benefit and evidence requirements, and (3) the lack of market-based incentives such as value-based reimbursement. Multiple approaches undertaken in parallel will be needed to address these challenges, as no single solution will be adequate or politically tenable.

Regulatory Policy

Over the past 20 years, several national expert panels have been commissioned to evaluate the regulatory environment for genetic tests in the United States, with each group consistently making recommendations for strengthening the system of federal oversight (Holtzman and Watson 1997; SACGHS 2008; SACGT 2000; SACGT 2001). Despite these findings, little has changed in how genetic tests are regulated, even in the face of dramatic changes in our understanding of the genetic basis of disease risk and drug response and the increasing technical sophistication of many genetic tests. There are ongoing legal challenges to the current system (e.g., in 2009, Genentech filed a Citizen's Petition urging the FDA to eliminate the current two-track system of approval and strengthen its oversight of laboratory-developed tests; GenomeWeb 2008), but from a practical standpoint, the FDA does not have the resources (monetary or human) to conduct reviews of all genetic tests.

At least for the near future, it appears unlikely that there will be a single path for regulatory approval of genetic tests. It is similarly doubtful that new regulations will require that product developers conduct studies to evaluate the clinical utility of a new genetic test. The most recent expert panel recommended the creation of a mandatory registry of all laboratory tests to report test-performance characteristics, with the details to be worked out by a group of stakeholders (although this had not been done, and the deadline for implementing this recommendation had passed, at the time of this writing) (SACGHS 2008).

We expect that an evidence gap will continue to exist for new tests that receive approval for marketing through FDA or CLIA but then fail to obtain positive coverage decisions from payers. This is because FDA evaluates tests in terms of whether they are "safe and effective," whereas payers typically apply the standard of "reasonable and necessary," the latter criterion being much more related to an improvement in health outcomes (Burken et al. 2009). Plans to ensure that there is a better "match" between the information needs of end users (such as clinicians and payers) and information developers will most likely occur through voluntary arrangements with a variety of financing mechanisms, rather than by regulatory fiat. An example of a collaborative network of relevant stakeholders for promoting the development and dissemination of better information supporting the use of genetic tests in clinical practice is that recently convened by the CDC and NIH (Khoury et al. 2009).

Fortunately, there are also ongoing private-sector efforts to help product developers and clinical researchers navigate this postregulatory evidence "divide" by providing guidance documents that describe how to design clinical studies that will provide decision makers (payers, clinicians, and patients) with reasonable confidence that a medical technology improves health outcomes. These "evidence guidance documents" are developed by a broad range of stakeholders convened by the Center for Medical Technology Policy (or CMTP, a private, nonprofit research and policy group) in an open, transparent forum and are intended to be technology-specific. Pharmacogenomics is an area of interest for the CMTP, which recently published guidance on the topic of gene expression profile tests for early-stage breast cancer (Goodman, Dickerson, and Wilson 2009).

Clinical Utility

The term "clinical utility" is relatively specific to the field of genetic testing, and confusion has arisen in the past because basic and clinical researchers in genetics/genomics, as well as health care professionals without subspecialty training in genetics, have approached the problem with their own notions of whether a test result is "clinically useful" (the plain-language interpretation of the phrase). At this point, there has been sufficient consensus within the expanded field of genetic testing for there to be agreement about the definition of clinical utility. Equally importantly, the definition appears to be meaningful to payers and clinicians, as many recent evidence reviews have highlighted how the lack of clinical utility data has led to noncoverage decisions or the inability to make practice guideline recommendations. The question then becomes, how should researchers sequence and prioritize studies to begin to build the evidence base to support claims of clinical utility? Given resource and time constraints, one practical tool is to first model the overall risks and benefits of testing versus usual care. Formal risk-benefit analyses provide a framework for gathering data from disparate sources (i.e., indirect evidence) and providing quantitative estimates of net benefit (Garrison, Towse, and Bresnahan 2007). The challenges of this approach are modeling complexity, concerns of bias, and the need for assumptions. Further research to identify best practices and to integrate risk-benefit analyses with traditional evidence-based processes are needed. Clearly there will never be sufficient resources to conduct all desired studies, but the use of modeling techniques such as the "value of information" approaches can help prioritize the research agenda for genetic tests by making explicit the benefits of further information to a decision maker (Rogowski, Grosse, and Khoury 2009).

The recent emphasis on comparative effectiveness research as part of health care reform efforts in the United States, as well as increased levels of federal funding for this category of research, increases the likelihood that academic researchers will conduct more clinical utility studies (Wilensky 2009). Personalized medicine and molecular diagnostics have been identified as priority areas of investigation by the Institute of Medicine and the NIH, which will lead to better information about whether these approaches are likely to improve health outcomes for patients as compared to current practice. Recent substantial federal investments in health information technology (both for patient care and for research purposes) will enable the conduct of comparative effectiveness studies at lower cost and with faster turnaround times.

Reimbursement

Historically, the diagnostics industry has not been required to conduct studies to demonstrate that use of a given laboratory test improves health outcomes. Industry leaders have balked at what they view as "raising the bar" in the technology-assessment arena, pointing out that many other health care technologies are not currently held to such a rigorous standard. Our view is that genetic-test developers' claims of delivering greater value to patients and the health care system need to be substantiated by evidence, and that the criteria for designing the appropriate models and/or studies have already been developed (Ramsey et al. 2006). Moreover, expectations being created from the growing federal investment in comparative effectiveness research will only increase pressures on PGx test developers to provide evidence of clinical utility and cost-effectiveness, so that clinicians, patients, and payers can make more-informed decisions about the overall value of genomics-guided interventions (Garber and Tunis 2009). However, the demand for demonstration of cost-effectiveness must be met with a system that rewards innovation at a price consistent with the value provided, rather than with the cost of production. Value-based pricing and coverage with evidence-development agreements are two examples of strategies that may incentivize companies to generate better evidence of product value.

Of critical importance is the need to increase researchers' and developers' awareness of the information needs and real-world constraints of payers and clinicians. A second and equally important point is the need to redesign the research enterprise to produce the desired information regarding both clinical utility and cost-effectiveness over time. As our examination of the evidence supporting the clinical integration of PGx testing for warfarin has revealed, there are critical data gaps that will continue to exist without changes in the current translational framework. In considering different policy options, our immediate goal should be to better connect the various stakeholders in the development phase of research, so that scarce resources can be focused on answering the most critical questions governing how genetic tests are integrated into clinical practice. The longer-term goals should be to view the process of translation more holistically and to begin to reshape the product-development and grant-funding processes in a way that optimizes the benefits of genomic medicine for public health.

REFERENCES

Anderson JL, Horne BD, Stevens SM, et al. (2007). Randomized trial of genotype-guided versus standard warfarin dosing in patients initiating oral anticoagulation. *Circulation*. 116(22):2563–2570.

Ansell J, Hirsh J, Hylek E, Jacobson A, Crowther M, Palareti G; American College of Chest Physicians. (2008). Pharmacology and management of the vitamin K antagonists: American College of Chest Physicians Evidence-Based Clinical Practice Guidelines (8th Edition). *Chest*. 133(6 Suppl):160S–198S.

[BCBS] BlueCross BlueShield Association. (2009). Technology evaluation center criteria. BlueCross BlueShield Association Web site. http://www.bcbs.com/blueresources/tec/tec-criteria.html. Updated 2009. Accessed November 21, 2010.

Burken MI, Wilson KS, Heller K, Pratt VM, Schoonmaker MM, Seifter E. (2009). The interface of Medicare coverage decision-making and emerging molecular-based laboratory testing. *Genet Med*. 11(4):225–231.

Caraco Y, Blotnick S, Muszkat M. (2008). *CYP2C9* genotype-guided warfarin prescribing enhances the efficacy and safety of anticoagulation: a prospective randomized controlled study. *Clin Pharmacol Ther*. 83(3):460–470.

ClinicalTrials.gov. (2010). Clarification of Optimal Anticoagulation Through Genetics (COAG). ClinicalTrials.gov Web site. http://clinicaltrials.gov/ct2/show/NCT00839657. Updated October 27, 2010. Accessed November 21, 2010.

[CMS] Centers for Medicare & Medicaid Services. (2009). *Proposed Decision Memo for Pharmacogenomic Testing for Warfarin Response (CAG-00400N)*. Washington, DC: Centers for Medicare & Medicaid Services, U.S. Dept of Health & Human Services.

Collins FS, McKusick VA. (2001). Implications of the Human Genome Project for medical science. *JAMA*. 285(5):540–544.

Davis JC, Furstenthal L, Desai AA, et al. (2009). The microeconomics of personalized medicine: today's challenge and tomorrow's promise. *Nat Rev Drug Discov*. 8(4): 279–286.

DNADiva. (2008). Genelex's CEO Medicare comments on warfarin DNA testing. A Forum for Improving Drug Safety [blog]. http://personalizedmedicineblog.com/2008/09/03/genelexs-ceo-medicare-comments-on-warfarin-dna-testing/. Updated September 3, 2008. Accessed November 20, 2010.

Eckman MH, Rosand J, Greenberg SM, Gage BF. (2009). Cost-effectiveness of using pharmacogenetic information in warfarin dosing for patients with nonvalvular atrial fibrillation. *Ann Intern Med*. 150(2):73–83.

Evans WE, Relling MV. (1999). Pharmacogenomics: translating functional genomics into rational therapeutics. *Science*. 286(5439):487–491.

[FDA] U.S. Food and Drug Administration. (2004). *Innovation or Stagnation: Challenge and Opportunity on the Critical Path to New Medical Products*. Washington, DC: U.S. Food and Drug Administration.

Flockhart DA, O'Kane D, Williams MS, et al.; ACMG Working Group on Pharmacogenetic Testing of CYP2C9, VKORC1 Alleles for Warfarin Use. (2008). Pharmacogenetic testing of CYP2C9 and VKORC1 alleles for warfarin. *Genet Med*. 10(2):139–150.

Garber AM. (2001). Evidence-based coverage policy. *Health Aff (Millwood)*. 20(5): 62–82.

Garber AM, Tunis SR. (2009). Does comparative-effectiveness research threaten personalized medicine? *N Engl J Med*. 360(19):1925–1927.

Garcia DA .(2008). Warfarin and pharmacogenomic testing: The case for restraint. *Clin Pharmacol Ther*. 84(3): 303–305.

Garrison LP Jr, Towse A, Bresnahan BW. (2007). Assessing a structured, quantitative health outcomes approach to drug risk-benefit analysis. *Health Aff (Millwood)*. 26(3):684–695.

GenomeWeb. (2008). Genentech files Citizen Petition urging FDA to regulate all lab-developed tests. GenomeWeb Daily News. http://www.genomeweb.com/genentech-files-citizen-petition-urging-fda-regulate-all-lab-developed-tests. Updated December 12, 2008. Accessed November 21, 2010.

Goodman S, Dickerson K, Wilson R. (2009). *Gene Expression Profile Tests for Early Stage Breast Cancer*. Baltimore, MD: Center for Medical Technology Policy.

Guyatt GH, Haynes RB, Jaeschke RZ, et al. (2000). Users' Guides to the Medical Literature: XXV. Evidence-based medicine: principles for applying the Users' Guides to patient care. Evidence-Based Medicine Working Group. *JAMA*. 284(10):1290–1296.

Holtzman NA, Watson MS, eds. (1997). *Promoting Safe and Effective Genetic Testing in the United States: Final Report of the Task Force on Genetic Testing.* Washington, DC: National Human Genome Research Institute, National Institutes of Health.

Hughes DA, Pirmohamed M. (2007). Warfarin pharmacogenetics: economic considerations. *Pharmacoeconomics.* 25(11):899–902.

[IWPC] International Warfarin Pharmacogenetics Consortium, Klein TE, Altman RB, et al. (2009). Estimation of the warfarin dose with clinical and pharmacogenetic data. *N Engl J Med.* 360(8):753–764.

Kangelaris KN, Bent S, Nussbaum RL, Garcia DA, Tice JA. (2009). Genetic testing before anticoagulation? A systematic review of pharmacogenetic dosing of warfarin. *J Gen Intern Med.* 24(5):656–664.

Khoury MJ, Gwinn M, Yoon PW, Dowling N, Moore CA, Bradley L. (2007). The continuum of translation research in genomic medicine: how can we accelerate the appropriate integration of human genome discoveries into health care and disease prevention? *Genet Med.* 9(10):665–674.

Khoury MJ, Feero WG, Reyes M, et al.; GAPPNet Planning Group. (2009). The genomic applications in practice and prevention network. *Genet Med.* 11(7): 488–494.

Kim MJ, Huang SM, Meyer UA, Rahman A, Lesko LJ. (2009). A regulatory science perspective on warfarin therapy: a pharmacogenetic opportunity. *J Clin Pharmacol.* 49(2):138–146.

Lesko LJ. (2008).The critical path of warfarin dosing: finding an optimal dosing strategy using pharmacogenetics. *Clin Pharmacol Ther.* 84(3):301–303. PMID 18714317.

Limdi NA, Veenstra DL. (2008). Warfarin pharmacogenetics. *Pharmacotherapy.* 28(9):1084–1097.

McWilliam A, Lutter R, Nardinelli C. (2006). *Health Care Savings from Personalizing Medicine Using Genetic Testing: The Case of Warfarin* [Working Paper 06–23]. Washington, DC: AEI-Brookings Joint Center for Regulatory Studies.

Patrick AR, Avorn J, Choudhry, NK (2009). Cost-effectiveness of genotype-guided Warfarin dosing for patients with atrial fibrillation. *Circ Cardiovasc Qual Outcomes.* 2(5): 429–436.

Pepe MS, Feng Z, Janes H, Bossuyt PM, Potter JD. (2008). Pivotal evaluation of the accuracy of a biomarker used for classification or prediction: standards for study design. *J Natl Cancer Inst.* 100(20):1432–1438.

[PhRMA] Pharmaceutical Research and Manufacturers of America. (2005). *What Goes into the Cost of Prescription Drugs?* Washington, DC: Pharmaceutical Research and Manufacturers of America.

Ramsey SD, Veenstra DL, Garrison LP Jr, et al. (2006). Toward evidence-based assessment for coverage and reimbursement of laboratory-based diagnostic and genetic tests. *Am J Manag Care.* 12(4):197–202.

Rogowski WH, Grosse SD, Khoury MJ. (2009). Challenges of translating genetic tests into clinical and public health practice. *Nat Rev Genet.* 10:489–495.

Rosove MH, Grody WW. (2009). Should we be applying warfarin pharmacogenetics to clinical practice? No, not now. *Ann Intern Med.* 151(4):270–273, W95.

[SACGHS] Secretary's Advisory Committee on Genetics, Health, and Society. (2008). *U.S. System of Oversight of Genetic Testing: A Response to the Charge of the Secretary of Health and Human Services.* Washington, DC, Secretary's Advisory Committee on Genetics, Health, and Society.

[SACGHS] Secretary's Advisory Committee on Genetics, Health, and Society. (2009). *The Integration of Genetic Technologies into Health Care and Public Health: A Progress Report and Future Directions of the Secretary's Advisory Committee on Genetics, Health and Society*. Washington, DC: Secretary's Advisory Committee on Genetics, Health and Society.

[SACGT] Secretary's Advisory Committee on Genetic Testing. (2000). *Enhancing the Oversight of Genetic Tests*. Washington, DC: Secretary's Advisory Committee on Genetic Testing.

[SACGT] Secretary's Advisory Committee on Genetic Testing. (2001). *Development of a Classification Methodology for Genetic Tests*. Washington, DC: Secretary's Advisory Committee on Genetic Testing.

Schwarz UI, Ritchie MD, Bradford Y, et al. (2008). Genetic determinants of response to warfarin during initial anticoagulation. *N Engl J Med*. 358(10):999–1008.

Swen JJ, Huizinga TW, Gelderblom H, et al. (2007). Translating pharmacogenomics: challenges on the road to the clinic. *PLoS Med*. 4(8):e209.

Tatsioni A, Zarin DA, Aronson N, et al. (2005). Challenges in systematic reviews of diagnostic technologies. *Ann Intern Med*. 142(12 Pt 2):1048–1055.

Teutsch SM, Bradley LA, Palomaki GE, et al.; EGAPP Working Group. (2009). The Evaluation of Genomic Applications in Practice and Prevention (EGAPP) Initiative: methods of the EGAPP Working Group. *Genet Med*. 11(1):3–14.

Veenstra DL. (2007). The cost-effectiveness of warfarin pharmacogenomics. *J Thromb Haemost*. 5(9):1974–1975.

Wilensky GR. (2009).The policies and politics of creating a comparative clinical effectiveness research center. *Health Aff (Millwood)*. 28(4):w719–w729.

Zerhouni EA. (2007). Translational research: moving discovery to practice. *Clin Pharmacol Ther*. 81(1):126–128.

6

The Power of Knowledge: How Carrier and Prenatal Screening Altered the Clinical Goals of Genetic Testing

NANCY PRESS, BENJAMIN S. WILFOND, MITZI MURRAY, AND WYLIE BURKE

Whether a research finding is ready to move into clinical care usually depends on whether the research has led to an application that helps improve health outcomes by preventing or curing disease, ameliorating symptoms, or guiding treatment options. Reproductive screening and testing is one of the primary applications of genetics; yet it can be seen as a departure from the prevailing paradigm with respect to when and why a research application is deemed ready for prime time. It is often claimed that the primary goal of genetic testing in pregnancy is to provide *information* that prospective parents might want. As a result, reproductive genetic testing has moved very rapidly through the stages of the translational pathway, its progress typically slowed only by questions of clinical validity (i.e., the ability of the test to accurately diagnose the condition it is intended to detect). This rapid translation of genetic tests in the reproductive arena has had a key role in institutionalizing the idea that genetic information is a worthwhile endpoint in itself, even when it does not point toward any evidence-based clinical action to be taken. It is our intention in this chapter to trace the development of this "information as an endpoint" view of genetics through the lens of reproductive testing and suggest that, in fact, it depends on a sleight of hand in which the primary purpose of reproductive genetic testing—the ability to terminate a pregnancy for a fetal anomaly—has been hidden.

A BRIEF HISTORY OF REPRODUCTIVE GENETIC SERVICES

Counseling parents about their risk of conceiving a child with a genetic disease was one of the earliest clinical applications of scientific knowledge about inheritance. In the first half of the twentieth century, medical geneticists could diagnose many rare conditions in infants and children based on their symptoms, and could counsel family members about the likelihood of a subsequent sibling being born with the same condition, based on knowledge about Mendelian patterns of inheritance. In the past 50 years, increasing knowledge and rapidly advancing technology have made it possible to identify women and couples at risk for having a child with a particular disability and, more recently, to directly diagnose disorders in the fetus during a pregnancy. Today, most women in the United States encounter some sort of genetic screening or testing during even a routine, low-risk pregnancy.

Genetic research has led to common reproductive health applications in two main ways. The first involves autosomal recessive disorders that are rare but occur more commonly in specific populations, such as Tay-Sachs disease (more common among Ashkenazi Jews) and cystic fibrosis (more common among Northern Europeans). These diseases are very serious, causing profound cognitive or physical problems; often they are fatal in childhood. Generally, they are caused by a mutation in a single gene. To have the disease, it is necessary to inherit one copy of the gene mutation from each parent. People who have a single copy of the mutated gene are called "carriers" and are not themselves ill. It is only when such an individual has a child with another carrier that the baby is at risk of inheriting the disease. Thus, until an affected baby is born, family history does not provide a clue to the risk of this disorder.

The second type of reproductive health application concerns conditions that occur somewhat more frequently but lack a well-defined at-risk population. The mechanism that leads to these disorders is not as well understood or as predictable: even the birth of a baby with the disorder does not strongly predict risk to a subsequent sibling. Examples here include neural tube defects, such as spina bifida, and chromosomal abnormalities, such as Down syndrome.

Testing for Autosomal Recessive Disorders: Tay-Sachs Disease

Tay-Sachs disease (TSD) is a paradigmatic case for understanding the evolution of reproductive genetic services. Babies born with this inherited metabolic disorder appear normal at birth. However, within about three to six months of age, babies with TSD begin to develop profound neurological problems including blindness, seizures, and severe muscle weakness (hypotonia). Most children with classic TSD die before four years of age because of pneumonia (Kaback 1999). While anyone can be a carrier of the disease, TSD is much more common among some population groups, especially those of Eastern European (Ashkenazi) Jewish descent. About 1 in 31 Jews in the United States are carriers of TSD, compared with about

1 in 250 people in the general population (Petersen et al. 1983). Initially, all that medical genetics could provide to parents was a diagnosis of TSD, some prognostic information, and, perhaps most importantly, information about the risk of recurrence of the disorder in a subsequent pregnancy (a risk of 1 in 4). In the hope that more complete knowledge about the disorder would yield approaches to prevention or treatment, geneticists sought to identify the specific biological causes of TSD.

An important breakthrough occurred in 1969 when scientists discovered the basic mechanism of TSD: a deficit in the functioning of an enzyme (beta-hexosaminidase A) leads to the accumulation of lipids in the nerve cells of the brain, premature death of those cells, and thus the reversal of development and early death of the young child (Okada and O'Brien 1969). The research meant that early diagnosis of TSD in newborns could be obtained by showing the absence of beta-hexosaminidase A enzymatic activity in cells. This important scientific discovery initially had limited relevance for at-risk parents. However, it suggested the possibility that the absence of enzymatic activity might also be detectable in the fetus during the pregnancy. Parents who had this information *before* the birth of a baby thus had the opportunity to decide whether to bring the pregnancy to term.

In a remarkable example of rapid translation of a scientific discovery into clinical use, prenatal screening for TSD moved in only five years from a faint possibility based on the biochemical discovery to a screening program in more than 50 U.S. cities, with a total of over 100,000 people tested (Kaback, Rimoin, and O'Brien 1977). Several factors were important, including advocacy on the part of scientist-clinicians who were deeply involved with families at risk for TSD, the prevalence of the condition in a self-identified community that was friendly to science and medicine, and the existence within that community of organizations that could reach a large proportion of the population and were willing to help coordinate early-testing programs. The extreme severity of the disorder also meant that motivation was high to avoid the birth of children who were certain to die within a few years of birth.

But disease severity is a factor in favor of screening only so long as there is some action to be taken following a positive screen. There was not then, nor is there now, a treatment for TSD. Thus, the only remedy is preventing births of affected fetuses. One interesting story that is often told about TSD screening involves the creation of Dor Yeshorim (Raz and Vizner 2008). This not-for-profit organization was begun in 1983 by a rabbi who had lost several children to TSD. Dor Yeshorim offers anonymous genetic screening to Jewish individuals to help avoid mate selections that could lead to the birth of children with TSD and, more recently, other autosomal recessive diseases that occur more commonly among Ashkenazi Jews (Leib 2005). Individuals are tested, preferably in high school, and given personal identification numbers. Any couple contemplating a serious relationship can then ask to have their status assessed by Dor Yeshorim. They will be told only whether they, as a couple, are at risk and should therefore not contemplate marrying each other. Since Orthodox Judaism generally opposes selective abortion, this method provides the possibility of preventing TSD without the need for pregnancy termination. Dor Yeshorim presents a clever and somewhat exotic solution; however, it is an approach with relevance to only the extremely small percentage of Jews who are ultra-Orthodox and are willing to have their marriages so arranged.

In the broader community, the success of the screening program for TSD depended not on premarital carrier testing leading to "safe" marriages, but on the ability of an already pregnant woman to have an amniocentesis and terminate the pregnancy if the fetus was shown to be affected. The development of TSD carrier screening and prenatal testing programs was explicitly motivated by the desire to provide carrier couples the opportunity to diagnose and terminate affected fetuses (Laberge et al. 2010). Furthermore, a retrospective study done in 2000 examined more than 1,400 couples at risk to have children with TSD between 1969 and 1999 and found that of the 628 affected fetuses detected, 609 (97%) were terminated (Kaback 2000).

Yet, in 1969, when TSD screening began, abortion was illegal in the United States. Although the study cited above reports terminations from even the earliest days of the program—obtained by having a physician certify that the termination was required for the health of the mother—the widespread use of prenatal screening for TSD in the Jewish community ultimately depended on the availability of legalized abortion.

Population-Based Screening for Non-Mendelian Disorders: Maternal Serum Alpha Fetoprotein and Multiple Marker Screening

The test that truly revolutionized prenatal diagnosis, however, was maternal serum alpha fetoprotein screening (MSAFP). MSAFP screening was developed to detect neural tube defects (NTDs) in the fetus. NTDs are anomalies of the fetal brain and spine that occur when the spinal column of the fetus does not close properly during development; NTDs cause a spectrum of problems and are among the most commonly reported serious birth defects. The most severe NTD, anencephaly, is generally fatal shortly after birth, while less severe forms are associated with widely varying degrees of paralysis and other impairments. Neither family history nor the age of the mother gives a clue to current pregnancies that might be at risk; more than 90% of cases occur in women with no prior risk factors (Brock and Sutcliffe 1972).

In the early 1970s, researchers in the United Kingdom found an association between NTDs and elevated levels of alpha fetoprotein (AFP), a substance produced by a developing fetus, in maternal blood (Brock, Bolton, and Monaghan 1973; Brock and Sutcliffe 1972). The link between AFP and NTDs had biological plausibility: AFP could leak into maternal blood through an open spinal column, and thus elevated levels of AFP in maternal serum could indicate a fetal spinal column that was not forming properly. Moreover, this finding suggested that a relatively inexpensive, safe blood test could be used to screen for NTDs. Ultrasound and amniocentesis could then be used for diagnosis. The combination of an easy blood test, a set of serious disorders for which no high-risk population subcategory could be delineated, and the legal availability of abortion led to a revolutionary suggestion: why not offer screening to *all* pregnant women?

This move was revolutionary because it would essentially define every pregnancy as "at risk," regardless of the age of the mother or the family's genetic history. In addition, translation to clinical application would require very considerable effort on the part of health care providers, as well as elevated costs for the health care system. Thus, such a large step was seen to require large-scale studies to corroborate the initial scientific findings and test feasibility. The first such investigation was the UK Collaborative Study, a retrospective analysis of almost 20,000 women that confirmed the clinical validity of MSAFP as a screen for NTDs (Wald et al. 1977). It was rapidly followed by prospective studies in other parts of the United Kingdom and Canada in which more than 95% of women in the study population were tested (Bennett, Gau, and Gau 1980; Gardner, Burton, and Johnson 1981) Many of these studies were considered so low risk that informed consent of participants was not required, which led researchers in the United States to conclude that MSAFP screening was not only effective in identifying pregnancies at risk for NTDs, but also highly acceptable to pregnant women. By the time later, smaller studies in the United States began to show some significantly lower rates of test acceptance by pregnant women and their physicians (Gardner, Burton, and Johnson 1981; Madlon-Kay et al. 1992), MSAFP had already been institutionalized.

Epidemiologist John McKinlay has observed that large-scale pilot studies can have a paradoxical effect: by introducing a sizable number of practitioners and patients to the technology being assessed, the pilot study may appear to actually demonstrate, rather than interrogate, the usefulness of the innovation and thus become a factor enhancing the move toward clinical translation (McKinlay 1982). In the case of MSAFP screening, the definitive moment of clinical integration took place in 1985, when lawyers for the American Congress of Obstetricians and Gynecologists (ACOG) issued a Legal Alert on the use of the test. Noting that more than 100,000 women in the United States had already undergone MSAFP testing, the Alert advised ACOG members that this level of use suggested that clinicians might be vulnerable to malpractice charges if they did *not* offer testing (ACOG 1985). This Legal Alert is considered to have been tantamount to making MSAFP screening a standard of care in prenatal medicine (Annas 1985).

But the scope of MSAFP screening was not to remain limited to the detection of NTDs. In 1984, data were reported on the connection between abnormally *low* levels of MSAFP and fetal trisomies such as Down syndrome (DS) (Merkatz et al. 1984). MSAFP was not a very good screen for DS; it had far lower detection rates than for NTDs. While it is unlikely that anyone would have suggested instituting an MSAFP screening program solely to test for DS, as an add-on to an already existing test, it seemed like a bonus.

KEY HISTORICAL FACTORS: A NEW PROFESSION AND NEW CLINICAL OPTIONS

To understand how prenatal genetic tests migrated from specialty clinics that provided diagnoses and information about recurrence risks for rare diseases into general practice and became a routinized part of standard prenatal care, we must consider three other historical developments. These are (1) the creation of a new profession—genetic counseling—in perfect contemporaneity with (2) the establishment of amniocentesis as a safe and available way to diagnose fetal anomalies, and (3) the legalization of abortion.

The Birth of Genetic Counseling

The story of the creation of the profession of genetic counseling has recently been told in detail by historian Alexandra Stern (2009). It begins in 1969 with Melissa Richter, a professor of psychology and biology at Sarah Lawrence College, as well as director of their Center for Continuing Education. Ms. Richter saw a way to meld advances in genetic knowledge with the professional needs of the educated, middle-class women who came to the Center for Continuing Education with a desire to rejoin the workforce. She envisioned a profession that would be ideal for such middle-class mothers, especially as genetics clinics generally ran only once a week in those days. Genetic counseling would be a profession that combined counseling skills and the ability to provide psychological support—much like social work—with the specialized knowledge needed to explain the complexities of genetic disorders to those affected or at risk for disease. Starting as a small, specialty niche, the profession would grow slowly, keeping pace with the growth of genetic knowledge of rare diseases. It seems likely that genetic counseling would have remained within the sphere of the once-a-week genetic clinic had two other things not occurred at about the same time.

The Expansion of Amniocentesis

Amniocentesis was already a known technology in the 1950s, but it was being used narrowly, primarily for testing of Rh incompatibility between pregnant women and their fetuses (a rare condition that can produce serious illness in the newborn but which can be monitored and addressed if found during pregnancy). In her discussion of the history of amniocentesis, Ruth Schwartz Cowan (1993) references the distinction between the developmental phase and the diffusion stage of any technology. She states, "[W]hen a technology is in development it is changing rapidly and being applied narrowly . . . [in diffusion] . . . it is coming into routine use, becoming embedded in . . . a social matrix" (p. 11). For amniocentesis, the diffusion stage began when it was demonstrated, in 1955, that fetal cells removed from the amniotic fluid could be examined under a microscope and that the total number of chromosomes, including the sex chromosomes, could be cultured and directly viewed—a process called *karyotyping*. The initial application of karyotyping was to identify male fetuses of women who carried serious genetic conditions, such as hemophilia and some forms of muscular dystrophy, on their X chromosome. These X-linked conditions are very serious and, at times, fatal for males. Although there was a 1960 report from Copenhagen of a pregnancy being terminated because the fetus was male and the mother was a carrier for hemophilia, abortion was not legal or easily obtained in the United States at that time, and the use of amniocentesis and karyotyping for sex identification remained limited.

However, interest in amniocentesis increased greatly when, in 1959, French geneticist Jerome Lejeune found the chromosomal cause of DS. Lejeune showed that DS occurred when there was an anomaly called Trisomy 21—that is, when there were three, instead of the normal two, copies of chromosome 21. DS is the most common cause of severe cognitive impairment for which the genetic underpinning is known. It does not "run in families" or in particular population groups, but the risk increases markedly with maternal age, especially for pregnant women older than 35. The possibility of using amniocentesis to screen for DS in these women increased clinical interest in amniocentesis, and several large studies were begun to ascertain the safety of the procedure. The stage was set for amniocentesis to diffuse into routine use in 1975, when reassuring results were released from the National Institute of Child Health and Human Development (NICHD) National Registry for Amniocentesis Study Group (NICHD National Registry for Amniocentesis Study Group 1976).

In the following four years, the number of amniocenteses performed in the United States increased tenfold. During this same period of time, there were several women who, although over 35, had not been offered amniocentesis and subsequently gave birth to children with DS; these women successfully sued their obstetricians for malpractice (Rogers 1982). In one of these cases, the family was awarded payment for the child's medical costs for life (Becker 1978).

The Legalization of Abortion in the United States

The success of these lawsuits depended on the argument that if the women had been offered amniocentesis, they would have been able to terminate the pregnancy and thus avoid the costs of raising a child with DS. This, in turn, depended on the quite separate fact that pregnancy termination had become legal in the United States by 1973. While the impetus for the legalization of abortion had little to do with its use in cases of severe birth anomalies diagnosed during pregnancy, the availability of abortion as a legal procedure was the final piece of the puzzle necessary for the routinization of prenatal diagnosis.

Thus, the very first cohort of genetic counselors started their clinical training at just the point where amniocentesis was being offered for serious and untreatable conditions to a fairly large group of women, and at the exact moment when termination of these pregnancies had become widely available and legal—a truly remarkable confluence of events.

"SHARED SILENCES AND COLLECTIVE FICTIONS" REVISITED

Scholars Nancy Press and Carole Browner examined the implementation of MSAFP screening in California in the early 1990s, and they discovered what they referred to as a silence around the topic of pregnancy termination (Press and Browner 1993). As a result of an explicit political decision, state-provided patient-education materials did not mention abortion; nor was the topic mentioned during Press and Browner's observations of prenatal appointments in which MSAFP screening was discussed (Press and Browner 1997). These omissions were mirrored in the reasons women gave for accepting MSAFP screening, which very rarely mentioned pregnancy termination, but reflected the things that were emphasized in the prenatal appointments and educational materials: reassurance, information, and emotional preparation in the case of a positive test result.

Press also pointed out that these same lacunae and coded language could be seen in the scientific research literature (Press 2000). There, the goals of MSAFP screening were briefly listed in the first, descriptive paragraphs of articles, which typically mentioned the value of screening to reassure the vast majority of women who would screen negative, provide information about the pregnancy, and allow time for parents to prepare emotionally. Also mentioned in this literature were the possibility of providing for special medical and delivery-room preparations for the birth of a child with an NTD, and the future possibility of in utero treatment. That is, the focus was on what might be seen as ancillary goals of population-based screening, in that these proposed benefits apply to statistically rare situations, involve information that could be found out in other ways, or describe hoped-for future situations. More importantly, it is extremely unlikely that the MSAFP test—or any other reproductive genetic screening—would have become routinized if these were really the major goals. Yet, the vast majority of women in the Press and Browner (1997) sample accepted testing, and the great majority of women receiving a positive test result in California during the period of their study did, in fact, terminate a pregnancy following a positive test result. Press and Browner (1997) commented that the shared silence around the topic of abortion may well have enhanced the acceptability of the test by creating a useful fiction about test purposes, which allowed women to more easily accept testing (followed by termination) than if the subject were raised directly during the initial offer of screening.

In the past 17 years, little appears to have changed, and the United States is not the only place where this disconnect can be observed. In a recent study that investigated the awareness of a group of pregnant women in France about the "decisional implications" of the ultrasound and biochemical screening for DS they had undergone or declined during their pregnancy, approximately half said they were unaware that such screening might lead to decision-making about the continuation of their pregnancy. Data from France during this time period (1992–2000), however, show that over 90% of fetuses with DS diagnosed prenatally were, in fact, terminated (De Vigan et al. 2005).

Nondirective Genetic Counseling and Information as an Endpoint

Thus, the avoidance of the fraught issue of abortion, and the substitution of the goals of reassurance and information provision, has had a profound effect on the way prenatal screening is perceived. And, for the minority of women and couples who receive prenatal screening results indicating an increased risk that their fetus will have a birth anomaly, the nondirective philosophy of genetic counseling has further shaped the view of genetic information as an endpoint.

The founder of the profession of genetic counseling, Melissa Richter, is reported to have personally believed that both individuals and society would be better off if the birth of children with disabilities were avoided (Stern 2009). But the burgeoning of the profession of genetic counseling did not occur until the Sarah Lawrence program came under the subsequent and formative leadership of Joan Marks. From that point, the philosophy of genetic counseling moved in a direction explicitly in opposition to any slight whiff of a eugenic view in which an end purpose of genetic counseling was to affect the composition of the population. Instead—and bolstered by the growing biomedical conversation about personal autonomy in medical decisions—nondirectiveness, value neutrality, and patient choice became bywords of genetic counseling. As the use of prenatal screening increased, and as the profession of genetic counseling grew along with it, the values of counseling solidified: these were seen in the provision of complete, technical information in a manner that fervently and intentionally steered clear of any preference for how the client might subsequently decide to use that information.

Much has been written about the difficulties of maintaining a completely nondirective stance in actual practice. However, the ideal of nondirective counseling, and the emphasis on the value of providing information, remains intact. Empirical support for this assertion is provided by data collected by authors of this chapter in a recent survey of the membership of the National Society of Genetic Counselors on the topic of prenatal screening for cystic fibrosis (unpublished data 2010). When asked, in an open-ended question, to define the primary purpose of this screening, the great majority of respondents gave answers that highlighted information provision. Conversely, when asked to indicate level of agreement with the statement, "The essential purpose for offering any prenatal screening for a fetal anomaly is to afford the opportunity of pregnancy termination for that anomaly," nearly three-quarters of respondents (73%) disagreed.

"Personalized Medicine" and the Value of Information

In the usual model of clinical screening, a potential test gets moved along the translational pathway into clinical care when there is a specific action to take following a positive or negative test result. However, the development of the emphasis on information as an endpoint, discussed above, has, we believe, helped shape a quite different view of genetics—one in which information is seen to be a worthwhile endpoint in itself—and this emphasis has moved quickly from reproductive genetics into the realm of primary care medicine.

Research on the genetics of common diseases has resulted in an increasing array of gene variants associated with moderately increased risk of heart disease, diabetes, cancer, and other major public health burdens. While few such findings connect a test result with a specific clinical action (for example, pharmacogenetic testing that indicates appropriate choices or doses of certain drugs; see Chapter 5), most of these scientifically interesting findings are very preliminary and do not suggest that the clinician manage a patient's care differently based on a test result. Nevertheless, interest in what is being termed "personalized medicine" is very high, with the term mentioned in the media as well as by funders and policy makers (Feero, Guttmacher, and Collins 2010; Khoury et al. 2009; Wikipedia contributors 2010).

While genetic information may not be linked to changes in treatment, claims have been made for an indirect health effect. The idea is that receiving information about one's personal genetic profile will have a particularly strong motivational value leading to behavior change. Thus, the genetic information is not intended to change *clinician* behavior but rather to increase *patient* compliance with standard and general information. For example, while every clinician would advise any patient who smoked cigarettes to stop doing so, some proponents of genetic testing assert that learning that one has a higher-than-average risk of becoming addicted to nicotine would have greater motivational power than a simple reminder about the health risks of smoking. Similar approaches involve using genetic information to motivate healthier diets. Perhaps because the usual suggested actions following testing—primarily involving changes in what might be termed "lifestyle"—do not require direct clinician involvement, testing for these genetic variations is increasingly being offered directly to consumers, often via the Internet. Although little harm seems likely to result from suggesting that one stop smoking or eat more broccoli, there is also no evidence to support the underlying claim that genetic information is especially motivating. Thus, although one prominent online genetics-service provider includes the tagline, "Confront risk early, prolong your health" (deCODEme 2010), research that has specifically examined the effect of genetic risk information on individual health behavior has found no effect, even for information based on robust genetic findings, such as for familial breast or colon cancers (deCODE genetics 2010).

Empiric challenges to the assumption of clinical benefit for personalized genetic information have, however, had little effect on slowing the exponentially increasing availability of such information. This may be because the view that genetic information has value in itself makes a demonstration of clinical benefit something of a side issue. Rather, information is offered based on the claim that individuals, as autonomous agents and decision makers, have a *right* to such information. For example, in marketing materials for a test to identify persons at increased risk for heart attacks, deCODE references "the patient's right to know" about all of his or her relevant risk factors.

In an intriguing article on just this point, science writer Denise Grady considers taking a test which can potentially identify genetic susceptibility to type 2 diabetes. As a member of a family with a positive history of diabetes, she found both pros and cons: "At first glance, testing doesn't seem to offer much to people like me. We already figure the deck is stacked against us .... Then again, if I tested positive, the threat would seem more real. Maybe I would eat less, exercise more, check my blood sugar more often ...." However, her overall feelings were that this choice should be hers: "Right now, I'm leaning toward having the test if it becomes available. I'm not sure what I'd do with the results or whether they would mean anything for my future. But I'd like the information, and the right to decide for myself whether to act on it" (Grady 2006, p. D5).

In short, she sees patient choice, rather than improved health outcome, as the criterion for test use. This view conflicts with conventional medical practice, but it is consistent with the emphasis on information as an endpoint and on autonomous patient choice that has emerged from reproductive genetic testing. Certainly this trend to privilege information provision cannot be laid entirely at the doorstep of reproductive genetics—especially as purveyed by the specialty of genetic counseling. But the great irony of the contribution of genetic counseling to this development is that, in fact, information is *not* the primary endpoint in reproductive genetics. That primary endpoint is decision-making about abortion. Yet, in obscuring that central point, one unintended legacy of reproductive genetic testing has been to aid what might be considered the precipitous and premature translation of genetic tests to clinical application.

REFERENCES

[ACOG] American College of Obstetrics and Gynecology. (1985). *Professional Liability Implications of AFP Tests. DPL Alert.* Washington, DC: American College of Obstetrics and Gynecology.

Annas GJ. (1985). Is a genetic screening test ready when the lawyers say it is? *Hastings Cent Rep*. 15:16–18.

Becker v. Schwartz, (1978) 46 NY 2d 401 - NY: Court of Appeals

Bennett MJ, Gau GS, Gau DW. (1980). Women's attitudes to screening for neural tube defects. *Br J Obstet Gynaecol*. 87(5):370–371.

Brock DJ, Sutcliffe RG. (1972). Alpha-fetoprotein in the antenatal diagnosis of anencephaly and spina bifida. *Lancet*. 2(7770):197–199.

Brock DJ, Bolton AE, Monaghan, JM. (1973). Prenatal diagnosis of anencephaly through maternal serum-alphafetoprotein measurement. *Lancet*. 9(7835):923–924.

Cowan, RS (1993) Aspects of the history of prenatal diagnosis. *Fetal Diagn Ther* 8(suppl 1): 10–17

deCODE genetics. (2010). Type 2 diabetes. deCODEhealth Web site. http://www.decodehealth.com/type_2_diabetes.php. Updated 2010. Accessed November 21, 2010.

deCODEme. (2010). Genes and health. deCODEme Web site. http://www.decodeme.com/genes-and-health. Updated February 18, 2010. Accessed November 21, 2010.

deVigan C, Khoshnood B, Lhomme A, Vodovar V, Goujard J, Goffinet F. (2005). Prevalence et diagnostic prenatal des malformations en population parisienne. *J Gynecol Obstet Biol Reprod.* 34(1):8–16.

Feero WG, Guttmacher AE, Collins FS. (2010). Genomic medicine—an updated primer. *N Engl J Med.* 362(21):2001–2011.

Gardner S, Burton BK, Johnson AM. (1981). Maternal serum alpha-fetoprotein screening: a report of the Forsyth County project. *Am J Obstet Gynecol.* 140(3):250–253.

Grady D. Genetic test for diabetes may gauge risk, but is risk worth knowing? NY Times National Edition, Aug 8, 2006, p. D5.

Kaback MM. (2000). Population-based genetic screening for reproductive counseling: the Tay-Sachs disease model. *Eur J Pediatr.* 159 Suppl 3:S192–S195.

Kaback MM, Rimoin DL, O'Brien JS, eds. (1977). *Tay-Sachs Disease: Screening and Prevention.* New York: Alan R. Liss.

Kaback MM. (1999). Hexosaminidase A deficiency. In: *GeneReviews* at GeneTests: Medical Genetics Information Resource [database online]. Seattle, WA: University of Washington. http://www.ncbi.nlm.nih.gov/books/NBK1218/. Updated May 19, 2006. Accessed November 25, 2010.

Khoury MJ, McBride CM, Schully SD, et al.; Centers for Disease Control and Prevention. (2009). The Scientific Foundation for personal genomics: recommendations from a National Institutes of Health-Centers for Disease Control and Prevention multidisciplinary workshop. *Genet Med.* 11(8):559–567.

Laberge AM, Watts C, Porter K, Burke W. (2010). Assessing the potential success of cystic fibrosis carrier screening: lessons learned from Tay-Sachs disease and beta-thalassemia. *Public Health Genomics.* 13(5):310–319.

Leib JR, Gollust SE, Hull SC, Wilfond BS. (2005). Carrier screening panels for Ashkenazi Jews: is more better? *Genet Med.* 7(3):185–190.

Madlon-Kay DJ, Reif C, Mersy DJ, Luxenberg MG. (1992) Maternal serum alpha-fetoprotein testing: physician experience and attitudes and their influence on patient acceptance. *J Fam Pract.* 35(4):395–400.

McKinlay JB. (1982). From "promising report" to "standard procedure": seven stages in the career of a medical innovation. In: Milbank JB, ed. *Technology and the Future of Health Care.* Milbank Readers, Vol. 8. Cambridge, MA: MIT Press: pp. 233–270.

Merkatz IR, Nitowsky HM, Macri JN, Johnson WE. (1984). An association between low maternal serum alpha-fetoprotein and fetal chromosomal abnormalities. *Am J Obstet Gynecol.* 148(7):886–894.

[NICHD] *Eunice Kennedy Shriver* National Institute of Child Health and Human Development; National Registry for Amniocentesis Study Group (1976). Midtrimester amniocentesis for prenatal diagnosis. Safety and accuracy. *JAMA.* 236(13):1471–1476.

[NHGRI] National Human Genome Research Institute. (2010). Learning about Tay-Sachs disease. National Human Genome Research Institute Web site. http://www.genome.gov/10001220. Accessed February 23, 2010.

Petersen GM, Rotter JI, Cantor RM, et al. (1983). The Tay-Sachs disease gene in North American Jewish populations: geographic variations and origin. *Am J Hum Genet.* 35:1258–1269.

Okada S, O'Brien JS. (1969). Tay-Sachs disease: generalized absence of a beta-D-N-acetylhexosaminidase component. *Science.* 165(3894):698–700.

Wikipedia contributors. (2010). Personalized medicine. Wikipedia, The Free Encyclopedia. http://en.wikipedia.org/wiki/Personalized_medicine. Accessed July 28, 2010.

Press N. (2000). Assessing the expressive character of prenatal testing: the choices made or the choices made available? In: Parens E, Asch A, eds. *Prenatal Genetic Testing and the Disabilities Rights Critique*. Washington, DC: Georgetown University Press: pp. 214–233.

Press NA, Browner CH. (1993). "Collective fictions": similarities in reasons for accepting MSAFP screening among women of diverse ethnic and social class backgrounds. *Fetal Diagn Ther*. 8(S1):97–106.

Press N, Browner CH. (1997). Why women say yes to prenatal diagnosis. *Soc Sci Med*. 45(7):979–989.

Raz AE, Vizner Y. (2008). Carrier matching and collective socialization in community genetics: Dor Yeshorim and the reinforcement of stigma. *Soc Sci Med*. 67(9): 1361–1369.

Rogers TD. (1982). Wrongful life and wrongful birth: medical malpractice in genetic counseling and prenatal testing. *South Carolina Law Review*. 713:33.

Stern AM. (2009). A quiet revolution: the birth of the genetic counselor at Sarah Lawrence College, 1969. *J Genet Counsel*. 18:1–11.

Wald N, Cuckle H, Brock JH, Peto R, Polani PE, Woodford FP. (1977). Maternal serum-alpha-fetoprotein measurement in antenatal screening for anencephaly and spina bifida in early pregnancy. Report of UK collaborative study on alpha-fetoprotein in relation to neural-tube defects. *Lancet*. 1(8026):1323–1332.

Commentary on the Development Phase of the Translational Cycle

SARA GOERING, SUZANNE HOLLAND, AND KELLY EDWARDS

In the development phase of the translational cycle, attention turns to assessing the value of new interventions, with an eye to creating clinical guidelines for their use. Concerns about patient safety, test efficacy, and clinical validity require development research that produces reliable evidence on which to base determinations about the utility of clinical use. The evidence produced may be in favor of moving a potential intervention forward to the creation of clinical guidelines, returning to basic research to explore related possibilities, or rejecting the test outright. Research at the development phase thus focuses on the difficult problems of ascertaining likelihood of beneficial health outcomes and setting appropriate levels of risk acceptance.

In the development phase, normative questions focus on determining what counts as reliable evidence of benefit (and for whom), as well as on what qualifies as a benefit (e.g., provision of information versus improved health outcomes). Furthermore, progress turns on figuring out appropriate incentives for the costly and difficult work of developing that evidence. For instance, as Deverka and Veenstra point out in Chapter 5, ascertaining the probability of clinical benefits for any particular health application relies on the development of an extensive evidence base. In the current context, that research is typically funded from the private sector, which aims at passing the evidentiary bar set by regulatory agencies (e.g., for test kits approved by the U.S. Food and Drug Administration [FDA], or for laboratory services approved by the Centers for Medicare and Medicaid Services) and then attracting a potential market. For genetic tests—currently the major health application produced by genomic research—the regulatory bar is low: the FDA does not require premarket review for most tests offered directly by laboratories. This combination of minimal regulation and interests in profit may push genetic tests onto the market before a sufficient evidence base has reliably demonstrated clinical benefit. Once a test is out, little incentive exists to return to the costly work of producing a solid evidence base about clinical benefit, in part because reimbursements for such tests typically are linked to cost rather than value.

The element of responsive justice that may be most relevant at this stage is *responsibility*. As an element of justice, responsibility focuses on the role of practitioners in identifying how their practices and the background political and economic context in which they occur may systematically shortchange certain groups and/or work against equity. Regardless of whether or not people actively claim injustice, scientists should therefore look to see how common research practices and the settings and circumstances in which research is conducted may result in significant limitations for others, particularly those already medically underserved or otherwise disadvantaged. Working to alleviate or at least mitigate these inequities is the role of the responsible researcher.

Chapter 5 demonstrates the problem of incentivizing the production of appropriate evidence by examining the case of warfarin, an anticoagulant drug for which appropriate dosing may be linked to particular genetic profiles. The relevant pharmacogenetic testing is available, but it has not been widely adopted in clinical practice because of scattered and somewhat conflicting evidence about its clinical utility. Tests originating in different private companies measure different types and numbers of associated polymorphisms. As a result, producing large-scale cost-benefit analyses that have any analytical validity is quite difficult; yet that is the standard most valued by organizations devising clinical guidelines for implementation. Diagnostic tests may be available, then, but unused, despite the likelihood that they would improve health outcomes for people at risk of thromboembolism.

Rethinking researcher responsibility would recommend attention to the *reasons* for the lack of a sufficient evidence base. Raising the regulatory bar would be one option for slowing the transition from bench science to market, but that tactic might have the result of dampening or even shutting down development research rather than encouraging more of it. Deverka and Veenstra point out difficulties with the incentive structure for such research. They hope that development researchers will be able to answer not only the question "Does it work better than standard care?" but also "Is it worth it?" In other words, will it be cost-effective within the health care system?

One possibility might be to rethink the incentive structure for development-phase researchers. For instance, as we mentioned in Chapter 1, philosopher Thomas Pogge and economist Aidan Hollis have argued that pharmaceutical companies must be convinced to focus their efforts less on "me too" lifestyle drugs that appeal to a relatively well-off population in the global North and more on large-scale disease burdens that predominantly affect the relatively impoverished global South (HIV/AIDS, malaria, etc.). To change the focus, they recommend that we restructure the incentive system, shifting reimbursement from cost to health care outcomes. Their proposed Health Impact Fund (IGH 2009) would provide an option for pharmaceutical companies. Companies that register their products with the fund would agree to sell their products at nearly cost, but they would then receive supplementary rewards based on the assessed global health impact of the product. Funding for the new incentive structure would come from participating national governments, each providing an estimated 0.03% of gross national product. Pogge and Hollis make moral as well as economic arguments in favor of the Health Impact Fund. Though the Health Impact Fund focuses on development-phase research broadly, its emphasis on offering rewards based on health value over profit might provide a reasonable model to encourage the creation of the kinds of evidence that Deverka and Veenstra call for in the realm of genomics development (Chapter 5).

Taking responsibility might also push researchers to confront the questionable but unacknowledged drivers of some of their development-phase projects. As Press et al. report in Chapter 6, genetic testing in the reproductive realm has expanded greatly in a relatively short period of time. Now nearly every pregnant woman who seeks prenatal care is offered screening and/or testing to detect fetal abnormalities. But such tests typically have no potential to improve the health outcome for the fetus, as there are no treatments or interventions to aid the fetus. Positive tests routinely result in pregnancy termination. Yet despite that fact, many women who undergo the testing report that they never thought of it as linked to the possibility of abortion; they simply took themselves to be gaining information, or abiding by standard prenatal practice. Clinicians and genetic counselors providing the testing similarly frame their role as one of procuring information for their patients.

Interestingly, this approach to reproductive testing, in which genetic information is construed as the endpoint and benefit of testing, has expanded into the realm of "personalized medicine," with genomic testing for a much wider array of conditions. Indeed, as Press et al. report, some commentators now talk of a *right* to genetic information. But responsible development-phase researchers might want to question whether every bit of information that could be offered *should* be offered, and whether the downside of the provision of such information has been adequately explored. Disability-rights critiques of standard prenatal testing practices, for instance, express concern about the devaluing of existing people with the impairments tested for, the lack of understanding about disability that sometimes motivates selective abortion, and the possibility of significant expansions in prenatal testing panels as additional markers for genetically linked conditions are identified (Parens and Asch 2000). Perhaps the point is that information is never simply information; as the history of genetic testing in the reproductive realm shows us, it is almost always connected to corresponding actions (or inactions), whether or not these are widely acknowledged. In thinking about expanding the genetic tests available for use, then, development researchers ought to be cognizant of and responsive to the ways in which their products may exacerbate inequalities or discrimination rather than ameliorating them.

REFERENCES

[IGH] Incentives for Global Health. (2009). The health impact fund: making new medicines accessible for all. Incentives for Global Health Web site. http://www.yale.edu/macmillan/igh/. Updated November 2009. Accessed November 23, 2009.

Parens E, Asch A. (2000). *Prenatal Testing and Disability Rights*. Washington, DC: Georgetown University Press.

7

Integrating Genetic Tests into Clinical Practice: The Role of Guidelines

ANNE-MARIE LABERGE AND WYLIE BURKE

Genetic testing is anticipated to be one of the most valuable products to follow from the completion of the Human Genome Project (Collins and McKusick 2001). Predictive tests that identify gene variants associated with disease risk and drug response (Couzin and Kaiser 2007) are expected to improve both prevention and treatment of disease (Collins et al. 2003). In addition, new tests are becoming available to diagnose rare genetic disorders and to inform would-be parents of their risk of having children with known genetic conditions (GeneTests 2010; Pollack 2010). However, it is not always easy to define just how emerging genetic tests should be used in clinical practice.

The speed with which gene discovery can lead to development of a new genetic test has arguably resulted in too much testing, or "premature translation," with genetic tests made widely available before their clinical value has been demonstrated definitively (Hunter, Khoury, and Drazen 2008; Janssens et al. 2008). At the same time, genetic tests, like other health care services, can also be underused, as when tests are "lost in translation" (Lenfant 2003), or fail to reach all of those who could benefit. Minority and low-income women, for example, are less likely to have testing for inherited risk of breast and ovarian cancer than other women (Huo and Olopade 2007; Shields, Burke, and Levy 2008).

Faced with a growing array of new tests, clinicians look for information and guidance on how best to use them. Clinical practice guidelines represent an important tool to help clinicians use genetic tests appropriately in caring for their patients. This chapter discusses the history of clinical practice guideline development and its limitations, explores how guideline development might work and be evaluated with respect to the coming expansion of available genetic tests, and argues for greater transparency about evidentiary standards in guideline development. While better practice guidelines alone cannot address every potential problem in the use of genetic tests, we believe they can be an important part of the solution.

CLINICAL PRACTICE GUIDELINES: HISTORY AND LIMITATIONS

Clinical practice guidelines originated in the 1970s from attempts to address unexplained variations in clinical practice, mostly occurring along geographic lines, that could not be explained by differences in disease prevalence (Edelson 2000). A 1990 report of the Institute of Medicine (Field and Lohr 1990) concluded that both overuse and underuse of clinical services occurred, and that clinical practice guidelines could improve both. Less overuse would decrease medical costs and exposure to iatrogenic complications by reducing inappropriate or unnecessary care (Field and Lohr), less underuse will increase costs (at least short term), but will improve access to necessary care. By defining appropriate care, guidelines provide a benchmark to detect overuse and underuse of services. More recently, the quest for rigor in the development of clinical practice guidelines has fueled the rise of evidence-based medicine (i.e., the use of strong scientific evidence to support clinical decisions). Guidelines can improve the provision of care by including recommendations about how services should be provided. In an ideal world, appropriately crafted evidence-based guidelines would standardize practice patterns, making medical practice less wasteful and more consistent with what is considered appropriate care (Ross 2000). Yet the appropriate level of care may vary based on patient population, or even individual patient preference. Guidelines are *guides* and can be used as a starting point for shared decision-making while discussing care with patients.

The reality, however, is more complicated. In some cases, there is inadequate evidence to evaluate the test. In others, the methods used to evaluate the evidence and create the guideline are questionable (or unknown). In addition, even when the evidence is abundant and trustworthy, experts from different specialties or health settings may interpret it differently, so that guidelines sometimes conflict. The evidence may vary for different patient populations. Guidelines may also fail to take into account barriers to health care such as the cost of tests or access to follow-up services.

Evidentiary Challenges in Guideline Development

A major problem for guideline developers is that relevant evidence is not always available or of high quality (Shaneyfelt and Centor 2009). Studies are evaluated in terms of their appropriateness for the clinical question at hand, the care with which they are designed and implemented, the applicability of their population and setting to other uses for the test, and the overall consistency of data. In some cases, evidence is available about the accuracy and validity of the test (i.e., the potential for the test to predict future risk correctly) but not about its clinical utility (i.e., whether it will help guide patient management or lead to improvement in health outcomes).

Guideline developers are interested in both the accuracy of a new test and its clinical utility. The scientific "gold standard" for answering this kind of question is a randomized clinical trial, in which half of the study participants are randomly assigned to receive the test and the other half, the "control" population, are not tested. The health outcomes of individuals in the two subgroups are then compared to determine whether the test made a difference. Such studies are expensive and can take a long time—and they may not provide definitive information. For example, the people participating in the study might be healthier or younger than the patients most likely to be tested, or the effect of the test might be difficult to measure, particularly if the goal is long-term disease prevention.

In light of the challenges posed by randomized clinical trials, many tests are evaluated *empirically* instead—that is, by observing testing outcomes in practice, without controls for comparison. If a test appears to improve health outcomes in clinical care, this evidence may be sufficient to inform practice guidelines, and such an approach may even be preferable to a randomized trial insofar as it measures test performance under actual clinical conditions. For small effects, however, this type of evidence may not be convincing. In addition, some of the measures researchers use to evaluate a test in a time-limited controlled research setting, such as changes in patient management (medication, etc.) in response to the test result, may not yet provide evidence of a health impact directly related to the use of the test. For example, a test to identify women with a genetic susceptibility to breast cancer might be evaluated by the number of high-risk women who are offered intensive breast cancer screening following a positive test result. The researchers' assumption is that these women will pursue screening and experience reduced mortality from breast cancer as a result. But note that this research approach does not provide evidence of actual benefit of the test itself, only the benefit presumed to come as a result of more intensive screening. In evaluating such studies, guideline developers need to consider whether the researchers' assumptions were reasonable.

As part of the larger effort to improve practice guidelines, tools have been developed to help evaluate the evidence about genetic tests in a systematic manner. One of them is the ACCE Model framework developed by the National Office of Public Health Genomics, of the Centers for Disease Control (CDC). Its name is an acronym of the four components evaluated: analytic validity (ability of the test to accurately identify the genetic variant); clinical validity (ability of the test to consistently and accurately detects or predicts the intermediate or final outcomes of interest); clinical utility; and ethical, legal, and social implications of the test (CDC 2010). In 2004, the CDC's Office of Genomics and Disease Prevention launched a follow-on project, the Evaluation of Genomic Applications in Prevention and

Practice (EGAPP). EGAPP's methods build on the ACCE framework to add contextual factors and targeted outcomes, including potential harms (Teutsch et al. 2009). The EGAPP approach is intended for evaluation of genetic tests with a wide population application, such as the evaluation of the use of tumor gene expression profiling to predict risk of breast cancer recurrence (EGAPP Working Group 2009; Teutsch et al. 2009). Tests currently eligible for EGAPP review include those used to guide medical intervention, identify individuals at risk for future disorders, or predict treatment response or adverse events. The quality of the evidence is assessed for each component, and the Working Group's recommendations (for, against, or, if insufficient evidence exists, none) are based on whether the evidence demonstrates the benefit of testing. The guidelines issued by EGAPP so far have been notable primarily for calling attention to the lack of evidence. For example, the EGAPP Working Group concluded that the evidence was inadequate to assess the analytic validity and the clinical utility of gene expression profiling to guide treatment options in women with breast cancer (Teutsch et al. 2009).

Limitations of Expert Opinion

In the absence of reliable scientific evidence for a given test, clinicians may turn to trusted colleagues for advice. Similarly, in the absence of reliable evidence, guideline developers depend almost entirely on expert opinion, either of the members of the panel or of well-known leaders in the field. But even lifelong experience as a world-renowned expert carries less weight than does evidence from a carefully designed study, because the expert can provide only the perspective of one individual in a single, unstandardized setting (Detsky 2006). Experts may have built-in biases, as well. For example, an expert may be more likely than the average clinician to see patients with the most serious complications, skewing his or her experience with a given condition toward more severe cases. Thus, if a condition is variable in its effects such that a person may have a mild or a severe case, the expert's judgments may tend to err on the side of overtreatment. Other biases may stem from disciplinary expertise: an expert surgeon may be more likely to favor surgical treatment, while an expert radiologist may tend to promote radiological procedures. This may help to explain why guidelines issued by organizations representing different specialties sometimes come to different conclusions about a particular test. In addition, financial conflicts of interest may be present; for example, ties with the pharmaceutical industry are not uncommon among panels creating guidelines for drug use, and they are not always clearly disclosed (Choudhry, Stelfox, and Detsky 2002).

Pitfalls in Process

Another area of variability is the methods and processes used to evaluate the available evidence, reach consensus around practice recommendations, determine the strength of recommendations, and promulgate guidelines. Guidelines are developed by many different organizations, ranging from government agencies (e.g., the U.S. Preventive Services Task Force [USPSTF]) to professional organizations (e.g., the American College of Medical Genetics), to private entities and interest groups.

Many government-sponsored groups have established methods for guideline development, and their approaches are typically quite robust. For example, the Agency for Healthcare Research and Quality (AHRQ) is the lead federal agency charged with improving the quality, safety, efficiency, and effectiveness of health care for all Americans (AHRQ 2010). To fulfill this mission, it is mandated to support the USPSTF, which uses a rigorous method to develop preventive-care recommendations that are based on a systematic review of the available evidence, taking into account quality and relevance (Barton et al. 2007; Harris et al. 2001). The ACCE Model Project and the EGAPP Working Group, described earlier, represent two other examples of guideline development methods developed within government agencies in the United States, but contrary to the USPSTF they do not stem from a government mandate. Other governmental agencies in Canada, Europe, and Australia have developed similar approaches, and an international consortium (the Grades of Recommendation Assessment, Development, and Evaluation [GRADE] Working Group) has developed a system for grading both the quality of evidence available for evaluating a given test or intervention and the strength of the ensuing recommendations (Atkins et al. 2004).

Few nongovernmental organizations, on the other hand, state explicitly how they develop guidelines, grade the quality of evidence available for a given test, or determine the strength of their recommendations. As described earlier, it is sometimes the case that insufficient evidence exists on which to base a recommendation, and expert opinion is the best source available. In other cases, however, guideline developers have omitted a systematic review of the evidence and instead relied on the judgment of the panel, perhaps reflecting long-standing medical tradition in viewing their personal experience as the most valid source of guidance (Tonelli 1999).

Instruments have been developed to evaluate practice guidelines, assessing the methodology used and the transparency of the guideline development process, including the interests and associations of panel members and the degree to which the recommendations were subjected to external review (AGREE 2001; Atkins et al. 2004; Cluzeau et al. 1999). Many existing guidelines are deficient by these measures (Manchikanti et al. 2008; Navarro Puerto et al. 2008; Delgado-Noguera et al. 2009).

GUIDELINES FOR GENETIC TESTS: NEW COMPLEXITIES

Historically, the focus of clinical genetics has been on establishing a diagnosis for patients with rare genetic conditions. As genetic and genomic research generates an increasing body of information about potential genetic risk factors, however, genetic testing serves a range of different purposes, including the prevention and treatment of disease, as well as reproductive decision-making. Just how guideline developers will address these purposes, and the value judgments they require, is an open question.

Traditional Testing Paradigm: Testing for a Single Gene

In the early days of medical genetics practice, tests were limited to diagnosing rare conditions for which treatment was often unavailable. In this setting, the goals of testing are to identify the condition so as to avoid further testing, sometimes to establish prognosis, and to inform family members about their risks. Importantly, such a diagnosis can often provide families with the opportunity for genetic testing to guide reproductive decision-making (Williams 2001). When it first became available, genetic testing for cystic fibrosis (CF) was performed to confirm a clinical diagnosis of CF in symptomatic children. Once the child's mutations were identified, relatives could be tested to find out if they were carriers and could use this information for reproductive decision-making. This traditional use of genetic testing has been expanded to the use of CF carrier screening in individuals with no affected family members, because of the high rate of carriers in the general population.

One of the most encouraging developments in medical genetics is that genetic conditions are increasingly amenable to treatment. In the case of CF, for example, life expectancy has increased from early childhood to the mid-30s over the past four decades. As a result, interest is shifting from carrier testing (to avoid the birth of affected children) to newborn screening, which enables the provision of early treatment to affected infants (Marshall and Campbell 2009).

Another condition for which genetic testing can lead to dramatic clinical benefit is multiple endocrine neoplasia type 2A (MEN 2A), a genetic disorder that occurs in about 1 in 30,000 people. MEN 2A results in a lifetime risk approaching 100% of developing medullary thyroid cancer, a type of cancer that is difficult to treat (Wiesner and Snow-Bailey 2005). Based on clinical logic—not on randomized trials—experts now offer prophylactic removal of the thyroid to individuals found to have the condition by genetic testing. An international group of clinical endocrinologists issued a guideline to describe this practice approach (Brandi et al. 2001), which represents a consensus statement of experts in the field. Subsequent observation suggests that outcomes have been improved for children at risk as a result of this approach (Piolat et al. 2006; Skinner et al. 2005).

Not surprisingly, as genetic testing offers more information to guide patient management, interest in evidence-based practice increases as well. Furthermore, with the proliferation of genetic tests—including tests related to common diseases such as cancer, heart disease, and diabetes—clinicians other than medical geneticists are becoming involved in the provision of genetic tests. *BRCA1/2* testing to identify women at increased risk for breast and ovarian cancer, for example, is often undertaken by oncologists. Evidence-based practice guidelines may play a particularly critical role in ensuring that the use of genetic tests by physicians other than genetic specialists is successful (Greendale and Pyeritz 2001).

Like MEN 2A testing, *BRCA1/2* testing leads to new prevention opportunities. *BRCA1/2* mutations are far more common than MEN 2A, with an estimated incidence of about 1 in 1,000 individuals (Petrucelli et al. 2007). Women who test positive for a known cancer-causing mutation have a lifetime risk of breast cancer of 85%. They are offered prophylactic surgery (removal of breasts or ovaries, or both) to reduce cancer risk. However, because these are radical approaches to risk reduction given the personal meaning of breasts and ovaries, much effort has been expended to find other options. Soon after testing became possible, an expert consensus panel convened by the Cancer Genetics Studies Consortium (organized by the National Human Genome Research Institute) suggested that women with these mutations should be offered early mammography screening and serum CA-125 testing to screen for ovarian cancer (Burke et al. 1997). The group also called attention to the uncertainty in these recommendations because of the lack of evidence on clinical outcomes. Subsequent research proved the importance of this caution. Multicenter prospective studies have now established a role for magnetic resonance imaging of the breast as an adjunct to mammography screening in high-risk women (Saslow et al. 2007). Conversely, screening trials have failed to demonstrate effective methods for ovarian cancer screening (Nossov et al. 2008), so that removal of the ovaries after completion of childbearing is now the intervention of choice for women with a high risk of ovarian cancer (Morgan et al. 2008).

New Genetic Testing Paradigm: Susceptibility Testing for Common Diseases Using Multiple Genes

The difficulty now faced in developing clinical practice guidelines is the increasing number of tests that provide risk information related to common diseases. Often many different genetic factors are associated with the same disease, each with a small effect on the total risk of developing disease. For example, about 20 different gene variants have shown some association with risk of type 2 diabetes—and, of course, risk is also affected by weight and diet (Wolfs et al. 2009). Even if a genetic test for diabetes proves to be a useful predictor of risk, it may not make much difference in terms of clinical management: the knowledge that a person's genetic risk of developing diabetes is slightly above normal would not change his or her physician's recommendations for preventive measures (healthy diet, exercise, etc.). Sorting out which genetic effects are both genuine and strong enough to affect clinical decisions requires large study populations and careful statistical methods (Ioannidis et al. 2008). Researchers are still working on how to combine information about different risk factors to reflect as well as possible the person's true underlying risk of disease. As they do so, they also need to consider what risk information would actually help to improve disease prevention.

These new genetic tests for common disease risk represent a paradigm that is notably different from traditional uses of genetic testing. Rather than diagnosing a specific genetic disorder, they identify a genetic risk that is often moderate, and clinically not much different from other common risk factors such as high blood pressure or cholesterol. Practice guidelines for genetic tests must address both emerging tests of this kind—such as a test for genetic susceptibility to diabetes—and traditional predictive genetic tests such as testing an individual with a family history of MEN 2A. Genetic susceptibility tests, as tools for prevention, will need to judged by their potential to reduce risk and improve health outcomes, taking into account that many new tests will provide risk information that is much less precise than that provided by tests for specific single-gene conditions, such as *BRCA* or MEN 2A testing. By contrast, tests for reproductive decision-making for single-gene disorders, such as CF carrier testing, are also judged by whether they provide information that is valued by prospective parents and useful for reproductive decision-making.

WHAT CAN GUIDELINES ACHIEVE IN PRACTICE?

When faced with new tests, clinicians look for information about how they should be used in the context of patient care. Clinical practice guidelines are often used as a reliable and easily accessible source of summarized information about appropriate care.

What Makes a Good Guideline?

To be accepted by clinicians, practice guidelines need to be seen as legitimate and reliable. A guideline's credibility can be grounded in substantive authority (i.e., evidence based) and procedural authority (i.e., through use of a well-accepted, credible process) (Ross 2000). The relationship between the evidence and the resulting recommendation is not always straightforward. Outcome data are often expressed in terms of odds or risks. To translate this information into a recommendation about appropriate actions, guideline developers have to make value judgments about the importance of the outcome and the potential risks associated with the intervention (Willems 2000). Unfortunately, such judgments are not often explicitly stated in the recommendations.

In the absence of strong scientific evidence, guidelines could still be considered authoritative if the guidelines are broadly disseminated and if their development process follows a defined method, discloses information about the selection process of the contributors, is open to the public, and accounts for patient preferences (Ross 2000). Different tests may require different guideline-development processes, depending on the nature of the evidence and the degree to which value judgments are required. Without transparency or a review process, the credibility of recommendations may be limited. This is especially important in areas where evidence is still sparse, ambiguous, or controversial, as is the case for many genetic tests. It has even been suggested that recommendations should acknowledge alternate viewpoints along with the majority opinion on controversial topics (Sniderman and Furberg 2009).

Good guidelines also provide clinicians with clear recommendations about how and when to use the test, and for which patients testing is appropriate. In some cases, however, guidelines may provide confusing recommendations. For example, in 2002 the College of American Pathologists issued recommendations on the use of genetic tests for factor V Leiden, a risk factor for blood clots. The recommendations differentiated between indications for which the test is "recommended," "controversial," "not recommended," and "can be considered" (Press et al. 2002). It is unclear how clinicians should make use of guidelines when ambiguous language (such as "controversial") is used in the formulation of recommendations, or when the distinction between categories is unclear (e.g., "controversial" versus "can be considered"). By contrast, the USPSTF issues four categories of recommendations based on the quality of the evidence and the risks and benefits of a particular service: "strongly recommends," "recommends," "makes no recommendation," or "recommends against." The USPSTF language is straightforward and easy to interpret, and it also expresses the panel's level of confidence in the recommendation. An additional category that is less helpful is used in the situation where the evidence is lacking: "The USPSTF concludes that the evidence is insufficient to recommend for or against" (Harris et al. 2001); while informative, this category provides no guidance on how clinicians should proceed.

Finally, guidelines reflect the standard of care at a point in time. Eventually, new evidence becomes available and guidelines need to be updated. The frequency with which guidelines are reviewed and updated varies from one organization to another. Some organizations schedule regular guideline reviews and updates, while others lack a defined plan. Depending on the topic, guidelines might become outdated fairly quickly if new evidence is compelling. The amount of research being done in genetics suggests that ensuring timely updates could be a challenge for guidelines on genetic tests and services.

What Influences the Use of Guidelines in Practice?

Although practice guidelines represent the official position of the group who issued them, the authors or the group they represent have no authority to enforce their application. Many barriers to clinician adherence to guidelines have been described; clinicians may be unaware of the existence of guidelines, be unfamiliar or disagree with their content, doubt the guidelines' ability to lead to the expected outcome, feel unable to overcome the inertia of previous practice, or face external barriers (time constraints, patient preferences, etc.) (Cabana et al. 1999). Clinicians may have difficulty choosing among available guidelines if information about methodology or process is limited, and they may be unwilling to abandon their clinical judgment in favor of what some call "cookbook medicine" (Timmermans and Mauck 2005). Guidelines' applicability to common clinical situations and to different types of patients also influences their uptake (McKinlay et al. 2007; Nuckols et al. 2007).

In the Netherlands, compliance with guidelines was studied among 61 general practitioners for 47 recommendations from 10 different guidelines issued by the Dutch College of General Practitioners (Grol et al. 1998). Three attributes reduced use of the guideline: (1) controversial recommendations not compatible with current values; (2) vague and nonspecific recommendations; and (3) recommendations demanding a change in routine care. Guidelines are more likely to influence practice if they are focused on realistic clinical situations, kept up to date, presented in concise and accessible formats, and linked to quality indicators and clinician support tools (Grol and van Weel 2009).

Two Illustrative Cases of Guideline Development

To illustrate the challenges involved in promulgating and ultimately improving practice guidelines, we will consider the development of two genetic testing guidelines: one for CF carrier testing, and one for genetic susceptibility to diabetes.

As described earlier, CF carrier tests generate information that can assist reproductive decision-making (Box 7-1). In 1997, a consensus panel convened by the National Institutes of Health (NIH) recommended that CF carrier testing be

---

Box 7-1

Cystic Fibrosis Carrier Screening

Cystic fibrosis (CF) is a genetic condition associated with progressive loss of lung function. Affected people often also have deficiencies in digestive enzymes, and they may suffer malnutrition without dietary supplements. Median life expectancy is now in the 30s, but childhood deaths still occur.

Genetics: CF is an *autosomal recessive* disease, meaning that an affected individual inherits a gene mutation from each parent; two mutations results in disease. A person with a single mutation is a CF carrier—healthy but with a 50% chance of passing the mutation on to a child. If both parents are carriers, they have a 25% chance of having a child with CF with each pregnancy. CF carriers are more common among people of European ancestry (about 1 in 25 individuals) than those of African, Asian, or Native American ancestry.

CF carrier screening: The American Congress of Obstetricians and Gynecologists issued practice guidelines for CF carrier screening in 2001. These recommend that women of European ancestry be *offered* CF carrier screening either before pregnancy or as part of prenatal care; they also recommend that CF carrier screening be *made available* to parents of other ethnic backgrounds. If a pregnant woman and her partner are both found to be CF carriers, they are to be offered prenatal testing to determine whether the fetus has CF. If the fetus is affected, pregnancy termination is an option.

---

offered to pregnant women (NIH 1997). The evidence the panel considered in making this recommendation included the accuracy of the test, which was able to correctly identify a very high proportion of CF carriers. In addition, the panel considered the results of other studies, which demonstrated low rates of CF carrier testing in nonobstetric settings (suggesting that offering the test in the obstetric setting would increase screening rates), and the effects of testing on reproductive decision-making (i.e., the high proportion of couples who chose prenatal diagnosis and pregnancy termination when both partners were found to be carriers).

Following the NIH consensus panel's recommendation, in 2001 the American Congress of Obstetricians and Gynecologists (ACOG) issued a guideline that recommended that CF carrier screening be offered to pregnant women of European descent and made available to pregnant women from other ethnic backgrounds (Menutti 2001). ACOG is the leading professional society for clinicians specializing in women's health care, with more than 52,000 members (ACOG 2010). The guideline emphasized the need for patient-information resources to educate patients about CF and the fact that people of European descent are more likely to be CF carriers, compared with other races. It is for that reason, and because mutation-detection rates are highest in populations of European descent, that ACOG targeted that population. ACOG developed special information leaflets and educational tools to help with guideline implementation (Mennuti 2001). The ACOG guideline also stressed the importance of nondirective counseling of couples at risk to have a child with CF. Although prior studies had indicated a high termination rate when prenatal diagnosis revealed that a fetus had CF, the ACOG guideline did not propose abortion rates as a metric for measuring the success of carrier screening programs. Whereas previous carrier screening programs for other recessive diseases had the aim of reducing disease incidence (see, for example, Chapter 6), CF carrier screening was framed as a service to provide couples an opportunity to make informed reproductive decisions (Laberge et al. 2009).

Two years after the guideline was released, 67.1% of obstetricians had thoroughly read or skimmed the guidelines, and 77.4% had changed their practice patterns because of the guidelines (Morgan et al. 2005). When asked about their practice patterns, 65.8% reported offering CF carrier screening to all their pregnant patients, although only 27.4% used all the criteria found in the guidelines (Morgan et al. 2004). The majority of survey respondents rated the following factors as being of at least moderate importance when considering whether to offer CF carrier screening: liability from not offering screening if the child was born with CF (77.2%), confidence in their ability to interpret or deal with a positive screening test (59.5%), and their level of familiarity with genetics and with CF (58.9%) (Morgan et al. 2004).

Although the guideline on CF carrier screening appears to have been endorsed by the majority of ACOG members, the guideline may not have achieved its full potential. The target audience for a clinical practice guideline is all clinicians who see patients with the condition of interest; but the CF screening guidelines are available freely only to ACOG members (ACOG 2010), despite the fact that many pregnant women are followed by family physicians. The language of the guideline was also less than ideal: many of the obstetricians who were surveyed found the distinction between "offering" CF carrier testing to women of European descent and "making it available" to others—a difference in language that was intended to signal the difference in carrier prevalence in different races—to be confusing, and therefore chose simply to offer the test to all pregnant women (Morgan et al. 2004).

Although the NIH consensus process and the ensuing ACOG guideline were informed by screening trials of carrier screening, they also reflected expert judgment about who should be offered testing, and how it should be offered. The guidelines are consistent with U.S. obstetrics practice, as discussed in Chapter 6, with its emphasis on the value of genetic information for reproductive decision-making. However, this approach is not uniformly accepted. In response to the same evidence, the Society of Obstetricians and Gynaecologists of Canada (SOGC) issued an opposite recommendation stating that "screening of all women during pregnancy for CF carrier status cannot be recommended at this time" (Wilson et al. 2002). The SOGC recommended CF testing in pregnancy only for individuals who may be at increased risk for CF due to considerations of family history or clinical manifestations. Although the evidence available to both groups was the same, the risks and benefits were valued differently. For the Canadian group, the cost of recommending screening for all pregnant women was a major factor in the decision not to make that recommendation. In the context of a single-payer system, the potential benefits of CF carrier screening did not justify the potential harms and the burden on the limited financial and human resources of the Canadian health care system (Wilson et al. 2002).

A different view about CF screening policy comes from the Cystic Fibrosis Foundation and CF advocates. The CF Foundation places an emphasis on newborn screening for CF, instead of CF carrier testing (CF Foundation 2007). All U.S. states now perform newborn screening for CF, with the goal of initiating therapy as early as possible in order to improve health outcomes of infants born with CF. From this perspective, avoidance of CF births is not the goal, making the detection of CF carriers, and the ACOG guideline, unnecessary. On the other hand, the implementation of CF carrier screening in Italy has already led to a reduction in the incidence of disease in some regions, changing the potential impact of CF newborn screening (Castellani et al. 2009). These competing views illustrate that value judgments are an inevitable part of developing guidelines and play a particularly important role in decision making about reproductive genetics.

Diabetes provides an example that illustrates another set of challenges to be faced in developing guidelines for emerging genetic tests. Recent research has identified multiple gene variants associated with increased risk for type 2 diabetes (Stolerman and Florez 2009). This form of diabetes typically has onset in middle age and is an important risk factor for heart disease, kidney failure, and blindness. Knowing about genetic risk could theoretically improve disease prevention (Box 7-2).

Box 7-2

Testing for Genetic Susceptibility to Diabetes

Type 2 diabetes mellitus (T2DM) is a condition in which the body develops resistance to insulin, resulting in poor ability to handle dietary sugar or other simple carbohydrates. Over time, excess blood sugar levels harm blood vessels and other body tissues, leading to complications such as heart disease, kidney failure, and blindness.

Genetics: An unhealthy diet, excess body weight, and low level of exercise all contribute to increased risk for T2DM. Risk is also higher among people who have close relatives with T2DM, and a number of gene variants associated with increased risk have been identified. The effect is usually small for any one gene variant; however, a person who inherited several gene variants associated with increased risk might have a substantially higher risk.

Testing for diabetes risk: A genetic risk panel testing for multiple gene variants associated with risk for T2DM could have the following potential benefits for people identified as having increased risk: (1) provide motivation for a healthier lifestyle; (2) guide physician recommendations about testing for blood sugar or for diabetes complications; and (3) help to justify more-aggressive cardiac prevention measures, such as drug treatment for elevated cholesterol.

In approaching a guideline concerning a test for gene variants associated with diabetes risk, the first question would clearly be about evidence of clinical benefit. The panel would want to know how effectively the test promoted disease prevention: do people who know they are at increased risk based on the test actually improve their diet or exercise program? A comparison to other risk-assessment methods would also be important: how well does the genetic test assess risk compared to other information clinicians might use, such as weight, family history, or fasting blood sugar testing?

Ideally, the guideline-development panel would gather all available evidence and then subject it to a systematic review, looking carefully at the quality of different studies. If such an exercise were undertaken now for type 2 diabetes, it would be disappointing: there is very little evidence for tests assessing diabetes susceptibility, and none so far to establish benefit. For example, two studies showed that a genetic risk panel provided little additional information about risk as compared to conventional risk measures (Meigs et al. 2008; Talmud et al. 2010).

Over time, however, the amount of evidence will increase. If genetic testing benefits are eventually documented, the panel will then need to make judgments about the level of benefit. For example, if 10% of people who received information about increased diabetes risk improved their diet and exercise pattern, would that be a sufficient outcome to justify the test? The answer would depend on many factors—the cost of the test, the resources involved in administering and interpreting it, the proportion of people at risk, the efficacy of the test compared to other strategies to improve lifestyle behaviors—and is likely to vary for different health care locations. For example, the value of information about one's genetic risk of developing diabetes may depend in part on the availability of the resources, skills, and knowledge patients need in order to take steps toward effective risk-reduction. Thus, the test might be least likely to provide benefit in a health system serving a population with limited access to sources of fresh foods and safe places to exercise, particularly if the rate of diabetes is high in the population. In this kind of setting, funds might be better spent on community-based efforts to increase opportunities for maintaining healthy lifestyles for the whole community. Conversely, in a community with a low prevalence of diabetes, genetic risk information might be valued as a guide to identify the individuals at higher risk and to inform their decisions about dietary or other preventive measures. In addition, in both settings, the value of testing would be influenced by the proportion of false-negative results—that is, genetic test results that are "normal" in individuals who really do have an increased risk for diabetes. Such results are likely in tests for genetic susceptibility to diabetes because genetic risk is only one contributor to overall risk. It would be important, therefore, to ensure that test results are not the only information used to guide patient management, and to find ways of combining genetic risk information with other sources of risk (diet, lifestyle, weight, etc.).

In putting together a guidelines panel to consider a genetic test for diabetes risk, the sponsoring organization would need to consider not only the evidence itself but also the potential conflicts of interests at different points in the guideline-development process: Which sources does the evidence come from (for example, is it industry-funded research, sponsored by a company that is developing the test)? Which populations were studied? How was the relevant evidence selected? Who are the stakeholders involved in the development of guidelines, and what are their interests?

CONCLUSION

Practice guidelines are useful yet imperfect tools for improving the translation of new scientific knowledge to improvements in health care, ideally leading to improved health outcomes. Some limitations are inevitable: no matter how good the evidence and the guideline-development process, some decisions about test use are a matter of judgment and prioritization, not evidence, and reasonable people might disagree. This is particularly the case when testing is strongly influenced by personal and social influences, as in the case of reproductive genetic testing. Guidelines panels should ideally provide a clear description of the reasoning behind their recommendations, so that clinicians and their patients can determine whether the recommendations are applicable.

However, other limitations are amenable to improvement. As discussed in Chapter 5, efforts to improve the evidence base for genetic tests are an urgent priority, as the full benefits of genomic knowledge will not be realized without adequate evidence to evaluate genetic tests. Equally important is a continuing effort to improve the quality of practice guidelines for genetic tests, with particular respect to (1) refinement of methods to evaluate evidence about genetic tests (Teutsch et al. 2009); (2) attention to process—ensuring that guidelines provide clear information about their procedures for evaluating evidence and for reaching consensus; disclosure of affiliations; and, to the greatest extent possible, avoidance of conflicts of interest; and (3) inclusion of plans for broad dissemination and periodic updates. Some have also suggested the need for routine inclusion of additional expertise in epidemiology, statistics, and health care policy on guideline-development committees (Sniderman and Furberg 2009). With a growing trend toward patient-centered outcomes, strategies are also required to include patient perspectives in the development of guidelines (Krahn and Naglie 2008). Engaging underserved communities in this effort could play an important role in moving practice guidelines toward systematic attention to health disparities.

The performance of a genetic test is based not only on its analytical and clinical validity, but also on the likelihood that its results will lead to improved health outcomes (Sanderson et al. 2005). This goal requires evaluation of the benefits and risks associated with testing, including related clinical interventions, their effectiveness, and their social consequences (Burke et al. 2002). Access to the care indicated by testing results is also important: improved health outcomes will occur only if effective preventive strategies or treatments are available for patients found to be at higher risk based on the genetic test result (Feero, Guttmacher, and Collins 2008).

In theory, the integration of genetic tests into practice would be facilitated by the development and dissemination of evidence-based guidelines (Khoury et al. 2007). In reality, evidence-based guidelines are often not yet possible for recently available genetic tests due to lack of evidence. In addressing the evidence deficit, attention will need to be given not only to health outcomes but also to the social and health care context in which tests will be provided, as well as to the priorities and preferences of those who will be tested.

REFERENCES

ACOG Web site. http://www.acog.org/. Accessed March 13, 2010.

[AGREE] The AGREE Collaboration. (2001). Appraisal of Guidelines for Research and Evaluation (AGREE) Instrument. http://www.agreecollaboration.org/pdf/agreeinstrumentfinal.pdf. Updated September 2001. Accessed May 29, 2007.

[AHRQ] Agency for Healthcare Research and Quality Web site. http://www.ahrq.gov/. Accessed August 2, 2010.

Atkins D, Best D, Briss PA, et al.; GRADE Working Group. (2004). Grading quality of evidence and strength of recommendations. *BMJ*. 328(7454):1490–1494.

Barton MB, Miller T, Wolff T, et al.; U.S. Preventive Services Task Force. (2007). How to read the new recommendation statement: methods update from the U.S. Preventive Services Task Force. *Ann Intern Med*. 147(2):123–127.

Brandi ML, Gagel RF, Angeli A, et al. (2001). Guidelines for diagnosis and therapy of MEN type 1 and type 2. *J Clin Endocrinol Metab.* 86(12):5658–5671.

Burke W, Daly M, Garber JE, et al.. (1997). Recommendations for follow-up care of individuals with an inherited predisposition to cancer. II. BRCA1 and BRCA2. Cancer Genetics Studies Consortium. I. 277:997–1003.

Burke W, Atkins D, Gwinn M, et al. (2002). Genetic test evaluation: information needs of clinicians, policy makers, and the public. *Am J Epidemiol.* 156:311–318.

Cabana MD, Rand CS, Powe NR, et al. (1999). Why don't physicians follow clinical practice guidelines? A framework for improvement. *JAMA*. 282(15):1458–1465.

Castellani C, Picci L, Tamanini A, Girardi P, Rizzotti P, Assael BM. (2009). Association between carrier screening and incidence of cystic fibrosis. *JAMA*. 302(23): 2573–2579.

[CDC] Centers for Disease Control and Prevention. Genomic translation: ACCE Model process for evaluating genetic tests. Centers for Disease Control and Prevention Web site. http://www.cdc.gov/genomics/gtesting/ACCE/index.htm. Updated September 29, 2010. Accessed November 23, 2010.

[CF Foundation] Cystic Fibrosis Foundation. Genetic carrier testing: overview. http://www.cff.org/AboutCF/Testing/GeneticCarrierTest/. Updated July 9, 2007. Accessed November 23, 2010.

Choudhry NK, Stelfox HT, Detsky AS. (2002). Relationships between authors of clinical practice guidelines and the pharmaceutical industry. *JAMA*. 287(5): 612–617.

Cluzeau FA, Littlejohns P, Grimshaw JM, Geder G, Moran SE. (1999). Development and application of a generic methodology to assess the quality of clinical guidelines. *Int J Qual Health Care*. 11(1):21–28.

Cochrane Collaboration Web site http://www.cochrane.org/about-us/evidence based health-care. Accessed February 4, 2010.

Collins FS, Green ED, Guttmacher AE, Guyer MS. (2003). A vision for the future of genomics research. *Nature*. 422(6934):835–847.

Collins FS, McKusick VA. (2001). Implications of the Human Genome Project for medical science. *JAMA*. 285(5):540–544.

Couzin J, Kaiser J. (2007). Genome-wide association. Closing the net on common disease genes. *Science*. 316(5826):820–822.

Delgado-Noguera M, Tort S, Bonfill X, Gich I, Alonso-Coello P. (2009). Quality assessment of clinical practice guidelines for the prevention and treatment of childhood overweight and obesity. *Eur J Pediatr*. 168(7):789–799.

Detsky AS. (2006). Sources of bias for authors of clinical practice guidelines. *CMAJ*. 175(9):1033, 1035.

Edelson PJ. (2000). Clinical practice guidelines: a historical perspective on their origins and significance. In: Boyle PJ, ed. *Getting Doctors to Listen: Ethics and Outcomes Data in Context*. Washington, DC: Georgetown University Press: pp. 153–163.

[EGAPP] Evaluation of Genomic Applications in Practice and Prevention (EGAPP) Working Group. (2009). Recommendations from the EGAPP Working Group: can tumor gene expression profiling improve outcomes in patients with breast cancer? *Genet Med*. 11(1):66–73.

Feero WG, Guttmacher AE, Collins FS. (2008). The genome gets personal—almost. *JAMA*. 299(11):1351–1352.

Field MJ, Lohr KN, eds.; Committee to Advise the Public Health Service on Clinical Practice Guidelines, Institute of Medicine. (1990). *Clinical Practice Guidelines: Directions for a New Program*. Washington, DC: National Academies Press.

GeneTests: Medical Genetics Information Resource (database online). Seattle, WA: University of Washington; 1993–2010. http://www.genetests.org. Accessed March 14, 2010.

Greendale K, Pyeritz RE. (2001). Empowering primary care health professionals in medical genetics: how soon? How fast? How far? *Am J Med Genet*. 106(3): 223–232.

Grol R, Dalhuijsen J, Thomas S, in't Veld C, Rutten G, Mokkink H. (1998). Attributes of clinical guidelines that influence use of guidelines in general practice: observational study. *BMJ*. 317:858–861.

Grol R, van Weel C. (2009). Getting a grip on guidelines: how to make them more relevant for practice. *Br J Gen Pract*. 59(562):e143–e144.

Harris RP, Helfand M, Woolf SH, et al.; Methods Work Group, Third U.S. Preventive Services Task Force. (2001). Current methods of the U.S. Preventive Services Task Force: a review of the process. *Am J Prev Med*. 20(3 Suppl):21–35.

Hunter DJ, Khoury MJ, Drazen JM. (2008). Letting the genome out of the bottle—will we get our wish? *N Engl J Med*. 358:105–107.

Huo D, Olopade OI. (2007). Genetic testing in diverse populations: are researchers doing enough to get out the correct message? *JAMA*. 298(24):2910–2911.

Ioannidis JP, Boffetta P, Little J, et al. (2008). Assessment of cumulative evidence on genetic associations: interim guidelines. *Int J Epidemiol*. 37(1):120–132.

Janssens AC, Gwinn M, Bradley LA, Oostra BA, van Duijn CM, Khoury MJ. (2008). A critical appraisal of the scientific basis of commercial genomic profiles used to assess health risks and personalize health interventions. *Am J Hum Genet*. 82(3): 593–599.

Khoury MJ, Gwinn M, Yoon PW, Dowling N, Moore CA, Bradley L. (2007). The continuum of translation research in genomic medicine: how can we accelerate the appropriate integration of human genome discoveries into health care and disease prevention? *Genet Med*. 9(10):665–674.

Krahn M, Naglie G. (2008). The next step in guideline development: incorporating patient preferences. *JAMA*. 300(4):436–438.

Laberge AM, Watts C, Porter K, Burke W. (2010). Assessing the potential success of cystic fibrosis carrier screening: lessons learned from Tay-Sachs disease and beta-thalassemia. *Public Health Genomics*. 13(5):310–319.

Lenfant C. (2003). Shattuck lecture—clinical research to clinical practice—lost in translation? *N Engl J Med*. 349(9):868–874.

Manchikanti L, Singh V, Helm S II, Trescot AM, Hirsch JA. (2008). A critical appraisal of 2007 American College of Occupational and Environmental Medicine (ACOEM) Practice Guidelines for Interventional Pain Management: an independent review utilizing AGREE, AMA, IOM, and other criteria. *Pain Physician*. 11(3):291–310.

Marshall BC, Campbell PW III. (2009). Improving the care of infants identified through cystic fibrosis newborn screening. *J Pediatr*. 155(6 Suppl):S71–S72.

McKinlay JB, Link CL, Freund KM, Marceau LD, O'Donnell AB, Lutfey KL. (2007). Sources of variation in physician adherence with clinical guidelines: results from a factorial experiment. *J Gen Intern Med*. 22(3):289–296.

Meigs JB, Shrader P, Sullivan LM, et al. (2008). Genotype score in addition to common risk factors for prediction of type 2 diabetes. *N Engl J Med.* 359(21):2208–2219. Erratum in: *N Engl J Med.* 2009;360(6):648.

Mennuti MT. (2001). Lights! Camera! Action! *Obstet Gynecol.* 98:539–541.

Morgan MA, Driscoll DA, Mennuti MT, Schulkin J. (2004). Practice patterns of obstetrician-gynecologists regarding preconception and prenatal screening for cystic fibrosis. *Genet Med.* 6:450–455.

Morgan MA, Driscoll DA, Zinberg S, Schulkin J, Mennuti MT. (2005). Impact of self-reported familiarity with guidelines for cystic fibrosis carrier screening. *Obstet Gynecol.* 105:1355–1361.

Morgan RJ Jr, Alvarez RD, Armstrong DK, et al.; National Comprehensive Cancer Network. (2008). Ovarian cancer. Clinical practice guidelines in oncology. *J Natl Compr Canc Netw.* 6(8):766–794.

Navarro Puerto MA, Ibarluzea IG, Ruiz OG, et al. (2008). Analysis of the quality of clinical practice guidelines on established ischemic stroke. *Int J Technol Assess Health Care.* 24(3):333–341.

[NIH] National Institutes of Health, Office of the Director. (1997). *Genetic Testing for Cystic Fibrosis.* NIH Consensus Statement. 15(4):1–37.

Nossov V, Amneus M, Su F, et al. (2008). The early detection of ovarian cancer: from traditional methods to proteomics. Can we really do better than serum CA-125? *Am J Obstet Gynecol.* 199(3):215–223.

Nuckols TK, Lim YW, Wynn BO, et al. (2008). Rigorous development does not ensure that guidelines are acceptable to a panel of knowledgeable providers. *J Gen Intern Med.* 23(1):37–44.

Petrucelli N, Daly MB, Bars Culver JO, Feldman GL. (2007). *BRCA1* and *BRCA2* hereditary breast/ovarian cancer. http://www.ncbi.nlm.nih.gov/bookshelf/br.fcgi?book=gene&part=brca1. Updated June 19, 2007. Accessed March 13, 2010.

Piolat C, Dyon JF, Sturm N, et al. (2006). Very early prophylactic thyroid surgery for infants with a mutation of the RET proto-oncogene at codon 634: evaluation of the implementation of international guidelines for MEN type 2 in a single centre. *Clin Endocrinol (Oxf).* 65(1):118–124.

Press RD, Bauer KA, Kujovich JL, Heit JA. (2002). Clinical utility of factor V leiden (R506Q) testing for the diagnosis and management of thromboembolic disorders. *Arch Pathol Lab Med.* 126:1304–1318.

Pollack A. (2010). Firm brings gene tests to masses. *New York Times.* January 29:B1.

Ross JW. (2000). Practice guidelines: texts in search of authority. In: Boyle PJ, ed. *Getting Doctors to Listen: Ethics and Outcomes Data in Context.* Washington, DC: Georgetown University Press: pp. 153–163.

Sanderson S, Zimmern R, Kroese M, Higgins J, Patch C, Emery J. (2005). How can the evaluation of genetic tests be enhanced? Lessons learned from the ACCE framework and evaluating genetic tests in the United Kingdom. *Genet Med.* 7(7): 495–500.

Saslow D, Boetes C, Burke W, et al.; American Cancer Society Breast Cancer Advisory Group. (2007). American Cancer Society guidelines for breast screening with MRI as an adjunct to mammography. *CA Cancer J Clin.* 57(2):75–89.

Shaneyfelt TM, Centor RM. (2009). Reassessment of clinical practice guidelines: go gently into that good night. *JAMA.* 301(8):868–869.

Shields AE, Burke W, Levy DE. (2008). Differential use of available genetic tests among primary care physicians in the United States: results of a national survey. *Genet Med.* 10(6):404–414.

Skinner MA, Moley JA, Dilley WG, Owzar K, Debenedetti MK, Wells SA Jr. (2005). Prophylactic thyroidectomy in multiple endocrine neoplasia type 2A. *N Engl J Med.* 353(11):1105–1113.

Sniderman AD, Furberg CD. (2009). Why guideline-making requires reform. *JAMA.* 301(4):429–431.

Stolerman ES, Florez JC. (2009). Genomics of type 2 diabetes mellitus: implications for the clinician. *Nat Rev Endocrinol.* 5(8):429–436.

Talmud PJ, Hingorani AD, Cooper JA, et al. (2010). Utility of genetic and non-genetic risk factors in prediction of type 2 diabetes: Whitehall II prospective cohort study. *BMJ.* 340:b4838.

Teutsch SM, Bradley LA, Palomaki GE, et al.; EGAPP Working Group. (2009). The Evaluation of Genomic Applications in Practice and Prevention (EGAPP) Initiative: methods of the EGAPP Working Group. *Genet Med.* 11(1):3–14.

Timmermans S, Mauck A. (2005). The promises and pitfalls of evidence-based medicine. *Health Aff (Millwood).* 24(1):18–28.

Tonelli MR. (1999). In defense of expert opinion. *Acad Med.* 74(11):1187–1192.

Wiesner GL, Snow-Bailey K. (2005). Multiple endocrine neoplasia type 2. http://www.ncbi.nlm.nih.gov/bookshelf/br.fcgi?book=gene&part=men2. Updated March 7, 2005. Accessed March 13, 2010.

Willems D. (2000). Outcomes, guidelines, and implementation in France, the Netherlands, and Great Britain. In: Boyle PJ, ed. *Getting Doctors to Listen: Ethics and Outcomes Data in Context.* Washington, DC: Georgetown University Press: pp. 153–163.

Williams MS. (2001). Genetics and managed care: policy statement of the American College of Medical Genetics. *Genet Med.* 3(6):430–435.

Wilson RD, Davies G, Desilets V, et al.; Society of Obstetricians and Gynaecologists of Canada. (2002). Cystic fibrosis carrier testing in pregnancy in Canada. *J Obstet Gynaecol Can.* 24(8):644–651.

Wolfs MG, Hofker MH, Wijmenga C, van Haeften TW. (2009). Type 2 diabetes mellitus: new genetic insights will lead to new therapeutics. *Curr Genomics.* 10(2):110–118.

8

Genomics and the Health Commons

NORA HENRIKSON AND WYLIE BURKE

In 1975, Henry Howard Hiatt, then dean of the Harvard School of Public Health, pioneered the idea of a medical commons, arguing that a society's health care resources are limited and, like any other finite resource, must be managed (Hiatt 1975). Failure to do so, he claimed, would lead to the "tragedy of the commons": the overuse and eventual degradation of common resources (Hardin 1968). Contrary to these warnings, the United States maintains a fragmented fee-for-service health care system, funded primarily through employers, that seeks to maximize individual choice and often rewards volume of service over quality, effectiveness, or equity of care (Jennings 2009).

This system is expensive. Health care costs accounted for about 17% of the U.S. gross domestic product in 2008 and are increasing at twice the rate of inflation (NCHC 2009). If these expenditures were associated with improvements in health outcomes, they might be worth the price. However, the United States ranks 46th among countries in life expectancy, and it falls short in most other health measures as well (Jennings 2009). At the same time, the number of uninsured people in the United States tops 40 million (Gilmer and Kronick 2009), and as much as one-third of medical spending in the United States may be for services of undetermined quality or effectiveness (Kilo and Larson 2009).

Health care applications based on genomic technology are rapidly entering into this fragmented system, accompanied by dramatic claims about the benefits they will bring to medical care (Feero, Guttmacher, and Collins 2008; Gammon 2008). Whether genomics can be leveraged to help address the inequity and inefficiency of U.S. health care delivery or will simply propagate them may depend on our willingness to adopt a health commons perspective. To do so will require a systematic process of decision making with attention to health care value and resources for disease prevention. Lacking a health commons perspective, we are unlikely to make efficient use of genomic discoveries or ensure that their benefits are justly distributed.

DEFINING A HEALTH COMMONS

Despite widespread agreement about the need to reform the U.S. health care system, Hiatt's concept of a medical commons receives little acknowledgement outside of academic commentary (Brennan and Reisman 2007; Cassel and Brennan 2007; Lederberg 1994; Michels 1994; Sennett and Wolfson 2006;). Even with passage and early implementation of the Affordable Care Act of 2010, strong political resistance to expanding government support of health care continues, and "socialized medicine" remains a term of contempt in much of U.S. discourse.

But accepting that health care resources are finite and crafting health policies to maximize health benefit for all strata of society do not require a national health care system. Rather, a commons perspective forces a critical look at both value and justice in health care delivery. It also requires high-quality evidence, to inform policies that ensure the best use of common resources and, therefore, the expansion from Hiatt's original focus on medical care to focus on a *health commons*, defined as *the collective societal resources available to spend on population health improvement, including public health, preventive and medical care, and translational research*—that is, a concept of the commons that incorporates research addressing the safe and efficient progress from bench science to applications in clinical medicine. As new genomic tests and services enter the health care system, a health commons perspective can provide a framework for assuring that they are used wisely and equitably.

Hiatt argued that the commons framework would allow policy makers to anticipate problems in health care delivery. In particular, he noted three significant threats to the commons, all of which are relevant to emerging genomic technologies:

1. *Services that pose conflicts between the interests of individuals and society.* These conflicts can take one of two forms: (1) effective but expensive interventions that will benefit very few individuals, such as universal screening to identify rare opportunities for health benefit; or (2) expensive interventions that provide marginal benefit to a large number of people, such as chemotherapy that extends life by a few weeks. In genetics, an example of the first kind of conflict is the expansion of newborn screening to include rare conditions that lack definitive treatment, and of the second, the rapidly emerging wave of innovative and high-cost biological cancer therapies based on genomic discovery.

2. *Services of no value or undetermined value.* Hiatt provided historical examples of interventions once practiced widely with insufficient evidence of their effectiveness, including lobotomy and complete dental extraction. These practices eventually ceased as evidence emerged of their ineffectiveness and harms, but only after thousands of people had been treated. An example of a genetics service that proved to lack value is newborn screening for histidinemia, a biochemical condition inherited as an autosomal recessive trait associated with increased levels of histidine. Newborn screening for this condition was initiated in the 1970s on the assumption that it represented a metabolic disorder, but population-based studies indicated this condition neither produces disease nor requires treatment, and screening was eventually stopped (Levy 2003). Considerable concern remains about genomic applications reaching medical practice before there is evidence that they improve health or are safe (Janssens 2008; Rogowski, Grosse, and Khoury 2009).
3. *Failure to implement known effective prevention that results in unnecessary drains on common resources.* Hiatt used examples of carelessly administered vaccination programs which resulted in resurgence of infectious disease, and, more comprehensively, how a lack of focus on social conditions such as poverty and malnutrition can perpetuate avoidable health burdens. A genetics example is the potential failure to derive benefit from newborn screening for treatable diseases if access to effective therapies, such as a phenylalanine low diet for phenylketonuria, is not assured to affected infants identified via screening. From a commons perspective, it is also important to measure genomic innovation in comparison to current best practice. A genetically based smoking cessation program, for example, should be compared to optimal nongenetic strategies.

Genomics and the Commons: Promise and Threat

The promise of personalized treatment and prevention is a leading theme in discussions about the potential of genomics to transform medicine and public health (Guttmacher and Collins 2005). From a commons perspective, we wonder: will genomic innovations be treated like other interventions that emerge in the U.S. health market, where access is based on coverage and personal means and expands only as the price drops? Or, in contrast, will coverage decisions about genomic innovations be based on evidence that they improve population health (and if so, how will we measure such improvements)? Will genomic innovation improve or contribute to the inefficiency, cost, and poor outcomes of the U.S. health care system?

There is likely no one answer to these questions, given the diversity of innovations based on genomic discovery. But one thing is sure: genomic innovations are rapidly entering the health care delivery system. For some clinical tasks, genomic technology offers clear improvement over traditional methods. For example, genomic tools now allow identification of many infectious disease agents, providing clinically relevant information more rapidly and with greater sensitivity than previous methods. Platforms for multiplex testing and expanded sequencing technology may also improve our ability to identify food-borne pathogens and viruses, which could help clinicians and public health officials respond to emerging infectious disease threats (Albuquerque et al. 2009; Liu 2008; Peters et al. 2004; Weile and Knabbe 2009) As an example, genomic studies of swine-origin 2009 A(H1N1) influenza provided timely and useful insights into the virus's evolution that have led directly to rapid implementation of surveillance programs and to vaccine development (Garten et al. 2009).

Similarly, DNA-based screening for variants associated with high cancer risk, such as *BRCA1* and *BRCA2* mutations, has allowed for early intervention and improved outcomes in people who test positive for the mutations (USPSTF 2005). Especially in oncology, genomic research is providing increasingly precise ways to classify disease, with often direct effect on treatment choices (Maciejewski, Tiu, and O'Keefe 2009). Tumor genetics is being used to identify people with breast cancer most likely to benefit from adjuvant chemotherapy, and more tools to enhance diagnostic and prognostic ability will continue to emerge (Oakman et al. 2009). Several pharmacogenetic tests, used to identify whether a person is likely to have severe side effects to a drug or not respond to it, also show promise in guiding drug treatment. In other cases, genomic discovery has led to the development of new therapies that can produce radical improvements in prognosis, such as treatments for chronic myelogenous leukemia and gastrointestinal stromal tumors (Ma and Adjei 2009).

However, medical technology can cause harm as well as benefit (Farley 2009; Kilo and Larson 2009), and new technologies do not always perform as expected (Califf 2004). We can predict with confidence that that some genetic applications will achieve their anticipated benefit—or perhaps even exceed it—but others will not, or will cause unanticipated harms that may outweigh their benefits. Evaluating health outcomes from the use of genomic technology is therefore important (see also Chapter 10). In this context, Hiatt's three threats to the commons are all relevant when considering new genomic health applications (see Table 8-1).

Conflict between Individual and Societal Interests

Genomic technologies that will benefit relatively few people, such as "orphan disease" treatments, or that will generate modest benefits at high cost (see Box 8-1), are perhaps the most obvious threat to a health commons. A particular issue for genetics is the cost of enzyme replacement therapy (ERT) for rare single-gene disorders that can be identified through genetic testing: growing numbers of effective therapies are available, but costs can exceed $100,000 per year (Correa Krug et al. 2009). A related threat arises from the skyrocketing costs of clinical cancer therapeutics, many of which now stem from genomic and related molecular research and provide relatively small incremental benefits to survival

Table 8-1. GENOMICS AND THREATS TO THE HEALTH COMMONS

Threat to commons[a]	**Genomics might aid management of commons by:**	**Genomics might threaten commons by:**	**Practices supporting commons-focused use of genomics**
Conflict between individual and society interests	Guiding targeted use of resources through identification of people/subgroups most likely to benefit/harm from intervention Simplifying or increasing efficiency of diagnosis Identifying fetal disease earlier in pregnancy	Exhausting resources on genomic screening for rare variants Exhausting resources on new treatments with small benefits for few Expanding reproductive genetics: resources required to cover more tests, support parents' decisions	Systematic and transparent process for decision making that includes societal input
Services of undetermined value (premature or accelerated translation)	Critically defining and analyzing clinical utility Investing resources in effectiveness research • Comparative effectiveness • Cost-effectiveness (value)	Diverting resources away from nongenomic topics that may provide more benefit to more people Spending resources on "information only" tests Producing cascade effects Exhausting resources on genomic screening before evidence of benefit	Systematic, evidence-based health technology assessment
Known preventive services not used, leading to preventable disease and expense for treatment downstream	Identifying people/subgroups most likely to benefit/harm from preventive care Leading to novel prevention applications	Diverting resources away from known effective interventions (social support; policies; vaccination programs, interventions to improve access to screening)	Resources devoted to health disparities; social determinants of health

[a] Hiatt HH. (1975). Protecting the medical commons: who is responsible? *N Engl J Med.* 293(5):235–241.

Box 8-1

BIOLOGICS

The term "biologic" refers to therapies that derive from living organisms (Ropp 1993), such as vaccines and blood products. However, genomic technology is leading several new biologic therapies, such as the following:

Factor VIII replacement for hemophilia. Hemophilia is a genetic disease occurring in about 1 in 5,000 to 10,000 males. Affected persons lack Factor VIII, a protein necessary for normal blood clotting, and therefore bleed excessively from minor injuries. Regular infusion of factor VIII can prevent complications, which is preferable to treating each bleeding episode. However, this treatment is expensive, costing up to $300,000 a year.

Enzyme replacement for Pompe disease. Pompe disease is a rare genetic disease affecting 1 in 140,000 to 150,000 people. Affected persons lack the enzyme acid alpha-glucosidase, and disease can take a mild form or the most severe, which results in muscle weakness and heart disease starting in the first month of life, with death occurring by age 1. A small number of patients have received enzyme-replacement therapy, and the majority have experienced significant improvement in life expectancy and function. The cost of treatment is about $300,000 per year.

Biologic cancer care. Emerging biologic therapies for cancer represent another innovation made possible by genomic technology. An example is cetuximab, a monoclonal antibody that inhibits epidermal growth factor receptor inhibitor. It is used in gastrointestinal and head and neck cancers, costs $10,000 a month, and on average provides a few extra months of life. Another, the drug pralatrexate, is used to treat lymphoma. Its costs more than cetuximab and has no effect on survival, only shrinking tumors.

The price of drugs such as cetuximab are defended by their makers, who cite the high cost of research and development and of drug pricing as a way to recoup their investment. Moves to make generic forms of biologic drugs are being vigorously fought in Washington.

The high cost of cancer therapy and other biologic treatments, when looked at from a health commons perspective, immediately raises a flag as a threat to common resources. Although patients considering these treatments may understandably have a different view, from a population perspective an extremely high-cost intervention that provides a modest health benefit could potentially exhaust common resources.

These therapies emerge directly from genomic innovations, and indirectly from public dollars invested in the pursuit of genomic knowledge. From a commons perspective, some difficult questions arise quickly and inevitably: What access is there for patients who cannot afford to pay for biologic therapy themselves? What is being neglected or not covered in order to pay for these therapies? Or are insurers obligated to cover these drugs at whatever price because there are no alternatives (Pollack 2009)?

(Meropol and Schulman 2007). The cost of new biologic therapies tends to be well above that of existing treatments, forcing difficult discussions of rationing within systems seeking to provide universal coverage (Kelly and Mir 2009).

For these emerging treatments, a system of societal deliberation is needed to define therapies that should be provided with common resources, as well as the circumstances under which they should be used. Some countries have adopted a systematic process for determining what is covered by common resources, with the United Kingdom being probably the most cited example (Drummond and Banta 2009). These systems inevitably face the "threshold question": what level of benefit per cost is appropriate for an intervention to be covered from common resources? Such systems have many critics but arguably provide the most equitable benefit for available resources (McCabe, Claxton, and Culyer 2008). In the United States, for example, where no deliberate allocation system exists, expensive therapies are de facto rationed by health insurance coverage and income status; in this setting, serious illness can disrupt both the ability to work and access to health coverage, and medical expenses can result in bankruptcy (Sered and Fernandopulle 2005).

A commons perspective thus foregrounds the need for a process for making difficult coverage decisions based on shared notions of legitimacy. The UK experience suggests that procedures can be developed for this purpose but, not surprisingly, contentious debates remain (Drummond and Banta 2009; Hanney et al. 2007). Political theorists Gutmann and Thompson (2005) suggest that our democratic traditions place four requirements on such deliberations: (1) they should provide readily accessible reasons for the resulting recommendations; (2) the recommendations should be justified by ethical arguments (for example, saving money in and of itself would not be sufficient); (3) the recommendations should be respectful of those who disagree; and (4) they should be revisable, acknowledging that new evidence may justify a change.

Likewise, a commons perspective argues for transparency in the decision-making process. It may be appropriate to cover some expensive therapies and not others; for example, ERT that provides a dramatic improvement in function and life expectancy arguably has a stronger claim to coverage from common resources than a similarly priced biologic therapy that extends life by a few weeks. The issue here is not the "right" decision, but the fair and open process—one that includes community members and other people or groups potentially affected by the decision—that leads to that decision.

Introduction of Services of Uncertain Value

Since Hiatt's analysis in 1975, the need for high-quality evidence of clinical effectiveness has become even more urgent. A call for translational research to improve "bench to bedside" research has been advocated (Zerhouni 2006), but evidence for many interventions is still lacking. In genomics, a discipline founded on the promise of revolutionary health benefit, premature translation of innovations into health care is a great risk. Personal genomic tests have been introduced in both the clinic and the direct-to-consumer market without evidence of their safety or impact on health outcomes, or even a clear definition of what the relevant health outcomes might be. Instead, arguments for clinical use are based on a largely untested assumption that risk information is inherently beneficial and that physicians may even have an obligation to provide it. For example, consider the information the test manufacturer deCODE genetics provides to physicians about "deCODE MI," a genetic test that can identify a 1.3-fold higher risk (a risk considered modestly increased at best) of myocardial infarction (MI):

> Irrespective of whether deCODEMI™ results move a given patient into another risk category, physicians are likely to have numerous patients with well known risk factors for coronary heart disease. Many patients need to do more to stop smoking, lose weight, get more exercise, improve their diet, or remember to take their blood lipid and/or blood pressure medication – and efforts to get them to do so are not always as successful as they should be. Additional, personalized information indicating which patients may also be at increased genetic risk of heart disease can add weight to this medical advice and its relevance. For patients, it can increase their incentive to implement and stick to lifestyle modification regimens and prescribed medications that can reduce their chances of getting MI. There is also a strong case to be made that it is the patient's right to know about all of his/her relevant risk factors. This may be especially relevant for those having difficulties adhering to diet and weight recommendations, and to the physician's role in reminding them of the importance of compliance with his/her recommendations (deCODE genetics: deCODEMI™ Clinical Use).

This example raises a number of questions about the feasibility of using of genetic risk information as a general guide to individualized health care, including the degree to which genetic risk information is preferable to other ways of assessing risk (if at all), the costs of incorporating genomic information reliably and safely into the medical record (Dean 2009), and the potential for error introduced by complex decision-making (Califf 2004). As DNA-based testing is expanded from single tests with a specific clinical focus to arrays of tests providing a genomic risk profile, many patients will receive information of uncertain clinical significance that may lead to a demand for additional screening, diagnostic, or other services. Such a "cascade effect" could present a drain on the health commons and subject a patient to risks from procedures or drug treatment (Deyo 2002; Kohane, Masys, and Altman 2006; Henrikson, Burke, and Veenstra 2008; McGuire and Burke 2008).

With genomics, the evaluation process must pay particular attention to tests that focus on providing information rather than on guiding health care. Although the genomic age has prompted discussion and reevaluation of how we define clinical benefit (Burke et al. 2002; Khoury 2003; Melzer et al. 2008), there is still no clear consensus on the proper use of common resources to pay for tests that provide "information only" (i.e., when the test results cannot be used to improve treatment or prevention decisions). For example, while an individual who has an *APOE*-e4 mutation associated with increased risk of Alzheimer disease may find knowledge of the mutation useful in planning for his or her future (Zick et al. 2005), it is not clear who should pay for tests of this type. From a health commons perspective, tests that "only" provide information may or may not be the best use of scarce resources when other tests that directly inform therapies might provide a more tangible health benefit.

Underuse of Known Preventive Services

A focus on testing for genetic risk without accompanying rigorous study of its effectiveness compared to other interventions could also divert resources away from nongenomic interventions, particularly those focused on prevention such as smoking cessation, dietary modification, or suicide prevention. This threat is perhaps of biggest concern for the medically underserved, who already experience disparities in access to screening and other preventive care. If genomic-based prevention is available only to some, it might in fact increase disparities. For example, coverage of genetic tests other than newborn screening is often optional under Medicaid (SACGHS 2006). Thus, despite an evidence-based consensus that genetic screening and follow-up are appropriate for women at high risk of developing hereditary breast and ovarian cancer, women who receive Medicaid may or may not be able to obtain coverage for these services, depending on where they live. From a commons perspective, it is appropriate to ask what the moral foundations of such a policy are. Why is an accepted prevention service not covered for all participants in a government program? What is the cost in health and personal finances to those for whom *BRCA* test coverage is denied?

When proven prevention measures are underused and significant health care disparities exist, genomic innovation is unlikely to help unless it has a notably strong effect—and few examples of effective genome-based prevention are known. In particular, the association between genetic risk knowledge and lifestyle changes such as improved diet, smoking cessation, and more regular exercise remains either unproven or ineffective (Carlsten and Burke 2006; Henrikson, Bowen, and Burke 2009).

However, when genomics leads to specific, tailored interventions or makes expensive prevention efforts more cost effective, it can make an important contribution—as is the case when genetic tests are used to identify people with a high risk of breast, ovarian, or colon cancer for individualized prevention efforts. These tests identify rare conditions and therefore apply to only a small proportion of the population—but potential benefits of genomics for prevention could extend well beyond such risk-based screening. For example, hypothetically, genomic analysis might lead to understanding why children raised on farms have lower risk of developing asthma. Benefits would not necessarily be based on genetic testing: the analysis could conceivably reveal gene–environment interactions that provide information about the effect of environmental triggers, or it could lead to the development of novel prevention that would be beneficial to all, such as development of a new immunization to reduce the risk of asthma. Without rigorous assessment, however, beneficial uses of genomics will be difficult to distinguish from others with unproven promise.

A commons perspective acknowledges scarce resources and the need to prioritize health applications and research. A balance between implementing proven therapies—genomically derived or not—and investing in genomic research with the potential to enhance prevention over time will be an important challenge, requiring careful deliberation and input from the public. Research that takes seriously the perspective of the underserved, acknowledges the common resources that support research, and is combined with emphasis on low-cost prevention and policies that address known social determinants of health may help navigate this difficult balance.

IMPLICATIONS

A commons perspective requires societal deliberation about the best use of resources. As genomic applications move from research and into clinical care, a careful comparison with nongenomic alternatives is needed to determine when these new technologies provide value. Similarly, the use of resources for genomic versus nongenomic *research* needs to be considered in terms of potential for health value. Decisions about which services should be covered with common resources and what research ought to have the highest priority require evidence as well, since effective deliberation will be difficult without good information about health outcomes. Genomics is part of this conversation because of its potential to lead to better health outcomes and improved management of health resources (e.g., by identifying which individuals are likely to benefit from or be harmed by a particular drug). As with all new health care technology, however, significant threats to the commons could occur if the emergence of genomic health care applications into clinical care is not managed.

Another related challenge is making appropriate decisions about the relative investment in basic genomic research versus applied efforts. Basic science generates new understanding of the biology of health and disease, and may sometimes lead to entirely unanticipated translational opportunities (Baltimore 1978). It is therefore an important part of the national investment in health research—but what proportion of research resources should it command?

While a robust basic science enterprise is a societal good, valuable in itself, greater investments in translational research and health care evaluation are needed. Recent federal investment in comparative effectiveness research (a new type of research that seeks to demonstrate how well different interventions work compared directly to each other; Sox and Greenfield 2009)–is one step in the right direction. This type of evidence can provide critical information for decision makers who need to make difficult choices about the relative value of health innovations in the context of limited resources. High-quality evidence can also inform other policy decisions, such as value-based insurance design, an insurance pricing model heavily dependent on evidence-based medicine where patient cost-sharing is adjusted based on the clinical value of services (see Box 8-2) (Chernew, Rosen, and Fendrick 2007).

However, evidence alone cannot determine policy. There is also a need for deliberative processes that ensure that decisions are made fairly, with appropriate stakeholder input; that disagreements are treated respectfully; and that decisions are revisited when new information is available. One such method has been called "just deliberation," where a deliberative, reciprocal process between citizens and decision makers encourages all parties to offer "mutually acceptable reasons to justify the laws and policies they adopt," guided by Gutmann and Thompson's (2005) criteria: that recommendations be accessible, moral, respectful, and revisable.

Although genomics has the potential to guide more-efficient use of common resources—such as in diagnosis, pathogen identification, or risk-based clinical management—the use of genomic information is an expensive enterprise that entails significant opportunity costs to the health care commons. A transparent process of societal deliberation about resource use, a strong investment in evidence-based guidelines, and dedicated attention to the social determinants of

---

Box 8-2

VALUE-BASED INSURANCE DESIGN

Some insurers are taking a new approach to designing patient cost-sharing structures that are based heavily on evidence-based standards of care. This approach is based on convincing evidence that even small out-of-pocket costs can provide a significant barrier to known valuable services, while covered services of unknown value, such as high-end expensive imaging, can easily be overused.

Value-based design characteristics can include reduced or removed copayments for preventive care or chronic disease medications, increased copayments for unproven tests, free access to substance abuse treatment services and community health advice, and financial incentives to participate in worksite wellness programs.

Evidence is still emerging about the long-term effectiveness and value of such up-front investments in removing financial barriers to known effective treatments—and such programs only affect people who already have insurance coverage. But these programs have a potential to serve as an example of programs that enact a health commons perspective—an explicit recognition of finite resources, the need to manage them, and the importance of encouraging use of known effective interventions over unknown, rapidly emerging ones—in policy approaches that may help manage resources and improve health at the same time.

---

health must help form the backbone for policies that use our limited resources most effectively.

REFERENCES

Albuquerque P, Mendes MV, Santos CL, Moradas-Ferreira P, Tavares F. (2009). DNA signature-based approaches for bacterial detection and identification. *Sci Total Environ.* 407(12):3641–3651.

Baltimore D. (1978). Limiting science: a biologist's perspective. *Daedalus.* 107(2): 37–45.

Brennan T, Reisman L (2007). Value-based insurance design and the next generation of consumer-driven healthcare. *Health Aff.* 26(2):w204–w207.

Burke W, Atkins D, Gwinn M, et al. (2002). Genetic test evaluation: information needs of clinicians, policy makers, and the public. *Am J Epidemiol.* 156(4):311–318.

Califf RM. (2004). Defining the balance of risk and benefit in the era of genomics and proteomics. *Health Aff (Millwood).* 23(1):77–87.

Carlsten C, Burke W (2006). Potential for genetics to promote public health: genetics research on smoking suggests caution about expectations. *JAMA.* 296(20):2480–2482.

Cassel CK, Brennan TE (2007). Managing medical resources—return to the commons? *JAMA.* 297(22):2518–2521.

Chernew ME, Rosen AB, Fendrick AM. (2007). Value-based insurance design. *Health Aff (Millwood).* 26(2):w195–w203.

Correa Krug B, Doederlein Schwartz IV, Lopes de Oliveira F, et al. (2009). The management of Gaucher disease in developing countries: a successful experience in southern Brazil [Epub ahead of print]. *Public Health Genomics.* doi:10.1159/000217793

Dean CE. (2009). Personalized medicine: boon or budget-buster? *Ann Pharmacother.* 43(5):958–962.

deCODE genetics: deCODEMI™ Clinical Use. http://www.decodehealth.com/myocardial_infarction.php?sp=86, accessed December 2010.

Deyo RA. (2002). Cascade effects of medical technology. *Annu Rev Public Health.* 23:23–44.

Drummond M, Banta D. (2009). Health technology assessment in the United Kingdom. *Int J Technol Assess Healthcare.* 25 Suppl 1:178–181.

Farley TA. (2009). Reforming healthcare or reforming health? *Am J Public Health.* 99(4):588–590.

Feero WG. Guttmacher AE, Collins FS. (2008). The genome gets personal—almost. *JAMA* 299(11):1351–1352.

Gammon K. (2008). Leroy Hood: look to the genome to rebuild healthcare. *Wired.* 16.10. http://www.wired.com/politics/law/magazine/16-10/sl_hood. Accessed November 24, 2010.

Garten RJ, Davis CT, Russel CA, et al. (2009). Antigenic and genetic characteristics of swine-origin 2009 A(H1N1) influenza viruses circulating in humans. *Science.* 325(5937):197–201.

Gilmer TP, Kronick RG. (2009). Hard times and health insurance: how many Americans will be uninsured by 2010? *Health Aff (Millwood).* 28(4):w573–w577.

Gutmann A, Thompson D. (2005). Just deliberation about healthcare. In: Danis M, Clancy C, Churchill LR, eds. *Ethical Dimensions of Health Policy*. New York, NY: Oxford University Press.

Guttmacher AE, Collins FS. (2005). Realizing the promise of genomics in biomedical research. *JAMA*. 294(11):1399–1402.

Hanney S, Buxton M, Green C, Coulson D, Raftery J. (2007). An assessment of the impact of the NHS Health Technology Assessment Programme. *Health Technol Assess*. 11(53):iii–iv, ix-xi, 1–180.

Hardin G. (1968). The tragedy of the commons. *Science*. 162(5364):1243–1248.

Henrikson NB, Bowen D, Burke W. (2009). Does genomic risk information motivate people to change their behavior? *Genome Med*. 1(4):37.

Henrikson NB, Burke W, Veenstra DL. (2008). Ancillary risk information and pharmacogenetic tests: social and policy implications. *Pharmacogenomics J*. 8(2):85–89.

Hiatt HH. (1975). Protecting the medical commons: who is responsible? *N Engl J Med*. 293(5):235–241.

Janssens A. (2008). Is the time right for translation research in genomics? *Eur J Epidemiol*. 23(11):707–710.

Jennings B. (2009). Liberty: free and equal. In: Crowley M, ed. *Connecting American Values with Health Reform*. Garrison, NY: The Hastings Center.

Kelly CJ, Mir FA. (2009). Economics of biological therapies. *BMJ*. 339:b3276.

Khoury MJ. (2003). Genetics and genomics in practice: the continuum from genetic disease to genetic information in health and disease. *Genet Med*. 5(4):261–268.

Kilo CM, Larson EB. (2009). Exploring the harmful effects of healthcare. *JAMA*. 302(1):89–91.

Kohane IS, Masys DR, Altman RB. (2006). The incidentalome: a threat to genomic medicine. *JAMA*. 296(2):212–215.

Lederberg J. (1994). The reform forecast for society's healthcare commons: heavy fog and hazardous driving conditions. *Ann N Y Acad Sci*. 729:175–7; discussion 188–96.

Levy HL. (2003). Lessons from the past—looking to the future. Newborn screening. *Pediatr Ann*. 32(8):505–508.

Liu YT. (2008). A technological update of molecular diagnostics for infectious diseases. *Infect Disord Drug Targets*. 8(3):183–188.

Ma WW, Adjei AA. (2009). Novel agents on the horizon for cancer therapy. *CA Cancer J Clin*. 59(2):111–137.

Maciejewski JP, Tiu RV, O'Keefe C. (2009). Application of array-based whole genome scanning technologies as a cytogenetic tool in haematological malignancies. *Br J Haematol*. 146(5):479–488.

McCabe C, Claxton K, Culyer AJ. (2008). The NICE cost-effectiveness threshold: what it is and what that means. *Pharmacoeconomics*. 26(9):733–744.

McGuire AL, Burke W. (2008). An unwelcome side effect of direct-to-consumer personal genome testing: raiding the medical commons. *JAMA*. 300(22):2669–2671.

Melzer D, Hogarth S, Liddell K, Ling T, Sanderson S, Zimmern RL. (2008). Genetic tests for common diseases: new insights, old concerns. *BMJ*. 336(7644):590–593.

Meropol NJ, Schulman KA. (2007). Cost of cancer care: issues and implications. *J Clin Oncol*. 25(2):180–186.

[NCHC] National Coalition on Health Care. (2009, September). *Fact sheet—cost*. Available at http://nchc.org/facts-resources/fact-sheet-cost. Accessed October, 2009.

Oakman C, Bessi S, Zafarana E, Galardi F, Biganzoli L, Di Leo A. (2009). Recent advances in systemic therapy: new diagnostics and biological predictors of outcome in early breast cancer. *Breast Cancer Res.* 11(2):205.

Peters RP, van Agtmael MA, Danner SA, Savelkoul PH, Vandenbroucke-Grauls CM. (2004). New developments in the diagnosis of bloodstream infections. *Lancet Infect Dis.* 4(12):751–760.

Pollack A. (2009). Questioning a $30,000-a-month cancer drug. *New York Times.* December 4:B1.

Rogowski WH, Grosse SD, Khoury MJ. (2009). Challenges of translating genetic tests into clinical and public health practice. *Nat Rev Genet.* 10(7):489–495.

Ropp KL. (1993). Just what is a biologic, anyway? *FDA Consum.* Vol 27.

SACGHS (2006). Coverage and reimbursement of genetic tests and services: report of the Secretary's Advisory Committee on Genetics, Health, and Society. Washington, DC, US Department of Health and Human Services.

Sared S, Fernandopulle RJ. (2005). *Uninsured in America: Life and Death in the Land of Opportunity.* Berkeley, CA: University of California Press.

Sox HC, Greenfield S. (2009). Comparative effectiveness research: a report from the Institute of Medicine. *Ann Intern Med.* 151(3):203–205.

[USPSTF] U.S. Preventive Services Task Force. (2005). Genetic risk assessment and *BRCA* mutation testing for breast and ovarian cancer susceptibility: recommendation statement. *Ann Intern Med.* 143(5):355–361.

Weile J, Knabbe C. (2009). Current applications and future trends of molecular diagnostics in clinical bacteriology. *Anal Bioanal* Chem. 394(3):731–742.

Zerhouni EA. (2006). Clinical research at a crossroads: the NIH roadmap. *J Investig Med.* 54(4):171–173.

Zick CD, Mathews CJ, Roberts JS, Cook-Deegan R, Pokorski RJ, Green RC; for REVEAL Study Group. (2005). Genetic testing for Alzheimer's disease and its impact on insurance purchasing behavior. *Health Aff (Millwood).* 24(2):483–490.

Commentary on the Delivery Phase of the Translational Cycle

SARA GOERING, SUZANNE HOLLAND, AND KELLY EDWARDS

At the delivery phase of the translational cycle, attention turns to how to get the most promising developments out into clinical practice. To do this, evidence regarding those developments must be incorporated into clinical guidelines, which must then be implemented into clinical practice. At this phase of research, issues about fair distribution are paramount.

In some cases, significant obstacles to fair delivery might warrant a return to the discovery or development phase, to reassess the specified modes of intervention. For instance, in Chapter 1 we relay the example of a vaccine that requires cold storage and multiple doses but is intended for treatment of a condition found mainly in patients in a geographic area without regular electricity or consistent access to health care. This vaccine might be reworked to avoid the need for cold storage, or to minimize the requirement of return visits, as a way to maximize the health benefits at the delivery phase. In other cases, genetic tests that have surged through the discovery and development phases and which are clinically valid might be held back from full implementation within the health care system if, for example, insufficient evidence exists to show clinical utility for patients (e.g, is there clinical utility in knowing one's *APOE*-e4 status—and possible disposition to develop Alzheimer disease?), or if concerns arise about the opportunity costs of pursuing the new test (e.g., if covering expensive tests for very rare disorders or low-risk conditions would necessitate a loss of basic preventive care for the majority).

Closely related to questions about distribution of benefits and burdens is the need to ensure that recognition is granted to the knowledge and concerns of medically underserved populations. Individuals from such groups might be well placed to inform clinicians and health researchers about barriers to the uptake of new technologies in their populations, and about ways to redesign delivery to avoid those obstacles. Their situated knowledge constitutes a form of expertise that is valuable but not traditionally acknowledged.

Of the genomic interventions that pass through the development stage, some will have very limited clinical significance, especially for marginalized populations. As Fullerton points out in Chapter 3, a lack of samples from individuals who share one's ancestry group might mean that otherwise valid clinical tests (e.g., about a genetic predisposition to diabetes, cancer, or heart disease) may deliver accurate results only for individuals in majority ancestry groups. For others who utilize the tests, the meaning of the results will be unclear, due to "variants of unknown significance." Scaling up for widespread delivery in such instances ignores the skewed potential for benefit, as well as the opportunity costs incurred, particularly where primary health care is underfunded or unevenly distributed. So, for example, making *BRCA* tests widely available will be unlikely to help many African American women at risk for breast cancer, or most women at risk for breast cancer, for that matter (see Chapter 10), and may detract from the push to ensure mammography screening and appropriate follow-up for all. If routine mammography for women above a certain age is more cost-effective and more useful for a wider population than *BRCA* testing for preventing breast cancer, then ensuring its widespread availability should be prioritized over pushing the uptake of *BRCA* testing in clinical practice. Finally, acknowledging the impact of market forces on what gets pushed forward from development to delivery—and rethinking relevant incentive structures to encourage more attention being paid to public health effectiveness rather than to profit (see Chapter 5)—is central to ensuring fair delivery of genomic interventions that can address health disparities.

As Chapter 8 points out, when we add relatively expensive genetic tests or therapies, often with limited or unknown clinical utility, into a fragmented system of health care provision such as the one currently operating in the United States, we are likely to exacerbate existing inequalities in health and in the opportunities that good health often makes possible. The authors' acknowledgement of a "health commons" recognizes that we have finite societal resources to put toward health promotion, so we are in need of better evidence of the clinical utility of many genetic tests/therapies before we can attempt to implement them on a broad scale. Some tests have met this standard. For instance, newborn screening can identify children who may benefit from early treatment for diseases such as phenylketonuria and sickle cell disease, and some pharmacogenetic tests have been shown to improve drug safety. Of course, even tests with sufficient evidence of clinical utility may disproportionately benefit some groups over others. As a society, then, we have reason to consider broad public benefits from the investment of those funds, rather than fully individualizing conceptions of benefit, and we need to involve the public in deliberations about the fair use of such funds. On this last point, the voices of the medically underserved must be heard if we are to fully appreciate what a fair distribution of the benefits of genomic intervention would be.

In a related point, as Laberge and Burke note in the conclusion to Chapter 7, expert panels that typically play a strong role in determining the implementation of practice guidelines often do not mandate any period of public discussion and deliberation, much less recommend representation from patients or medically underserved groups. Yet such stakeholders may have relevant information about how the appropriate genetic tests could be best utilized by patients. Taking seriously the needs and interests of marginalized and medically underserved populations will require changes in the process of developing and evaluating clinical guidelines, as well as a responsibility on the part of traditionally recognized experts to ensure that these voices are heard.

At the delivery phase of the translational pathway, responsive justice highlights the importance of achieving fair distribution by recognizing nondominant voices and including them in the decision-making process about what health policies to create and how to distribute access to those benefits equitably. Their inclusion at this stage has benefits in terms of both understanding barriers to delivery and achieving fair delivery. By highlighting barriers to delivery and reporting on the reasons that underserved individuals might fail to make use of services that are clinically available, they can contribute significant information to the bigger picture of how delivery works, and where (and why) it fails. The available evidence points to a wide range of barriers—including competing priorities—that may be substantial for people concerned about adequate food or shelter, the location of services, costs, and the lack of culturally sensitive health care providers (Smedley, Stith, and Nelson 2002). For any given health care service, some of these factors may be more important than others, or they might be amenable to specific strategies to overcome them. Consideration of such factors will assure that clinical guidelines will be more likely to achieve health benefit in an equitable way.

Broader inclusion of the perspectives of medically underserved groups should also proffer ethical benefits. A health commons view could be read as a majoritarian move—one that will likely invest less in "orphan diseases" and perhaps also in diseases of concern to minority populations, because treatments for these offer less widely distributed benefits. However, in the health commons view proposed by Henrikson and Burke, populations that have historically been oppressed and impoverished should get access to basic preventive measures and routine treatments that they have previously been denied, as well as access to genetic innovations that prove to be more efficient in achieving traditional health goals. So, for instance, groups that have limited access to dementia care might be prioritized to receive that care over the implementation of a genetic susceptibility test for *APOE*-e4, associated with Alzheimer disease, which provides information but has no clear clinical utility. Relatively well-off individuals may desire that information and be willing to pay for it, but a health commons approach would question any attempt to have it covered by insurance. Responsive justice demands that genomic innovations be delivered fairly, but that they be delivered only if they don't distract attention and funding away from the provision of existing therapies that have the potential for a much broader public health impact.

REFERENCES

Smedley BD, Stith AY, Nelson AR, eds. (2002). *Unequal Treatment: Confronting Racial and Ethnic Disparities in Health Care*. Washington, DC: National Academies Press.

9

The Role of Advocacy in Newborn Screening

CATHARINE RILEY AND CAROLYN WATTS

Though advocates play a role in nearly every phase of the translational pathway, relatively little advocacy work has been done in relation to outcomes research. In a way, this makes sense: advocates typically direct their limited resources and social capital to areas of research that are of most concern to them, and to those in which they can have the greatest influence with the least expenditure of resources. Historically, advocates for research on genetic conditions have focused on disease identification, screening and diagnostic testing methodologies, and increased understanding of a disease so treatments and interventions can be developed. Less time and energy has been devoted to short- and long-term *evaluation* of the health outcomes delivered by new interventions. Genetic-disease advocates may presume that improved health will follow once treatments are identified and delivered. However, as described in Chapter 10, not all interventions deliver the health benefit promised or hoped for.

In this chapter, we delve into the history of newborn screening with a particular focus on the role of advocacy in shaping policy. Newborn screening, a public health program designed to screen all infants for a set of rare congenital and/or genetic conditions, is an interesting and complex case. The translation of newborn screening into clinical practice depends on decisions that are made within state-funded public health programs, often by elected officials with input from experts, government staff, and community advocates. As such, the history of newborn screening, a "gold standard" in the realm of public health genetics, demonstrates how advocacy can influence the uptake of new discoveries: diseases for which to screen, novel technologies and testing methods, and innovative interventions and treatments.

After reviewing the history of newborn screening and the role of advocates in securing its delivery, we turn to the relative lack of advocacy for outcomes research related to newborn screening. We consider how the recent push for significant expansions of newborn screening—made possible through technological advances—might be influenced by the kind of evidence that could be provided by newborn-screening-related outcomes research. In our discussion, we highlight the crucial role advocacy has played in advancing the science and practice of newborn screening, and we highlight the tensions between differently placed advocates and point to the importance of recognizing which advocates' voices are heard and which tend to be silent.

NEWBORN SCREENING, THE BEGINNING

Newborn screening would not be what it is today were it not for the discovery of phenylketonuria (PKU) and the subsequent identification of treatment options and screening mechanisms. Individuals who have PKU cannot break down phenylalanine, an amino acid most commonly found in protein. As a result, phenylalanine builds up in the body and results in severe cognitive impairment if left untreated (Kahler and Fahey 2003). The history of newborn screening for PKU demonstrates the important role that individual advocates and advocacy groups can play in urging researchers not only to identify diseases, but also to develop interventions and treatments to improve health outcomes. This case provides an interesting backdrop against which to discuss the limited role of advocacy in pushing for evaluation of long-term health outcomes.

Between the dawn of the Mendelian heredity theory in the 1910s and recognition of the field of molecular biology in the 1930s, Garrod, a British physician, applied Mendelian heredity theory to "inborn errors of metabolism," a concept that he pioneered (Guthrie, 1989; Scriver 2001). He concluded that genetic differences in metabolic pathways could lead to different health outcomes in individuals. In 1934, Dr. Asbjörn Fölling, a Norwegian doctor determined to understand the cause of mental retardation in two siblings who were his patients, discovered that some children with mental retardation have unusually high levels of phenylpyruvic acid in their urine compared to the average child. It is worth noting here that the mother of these siblings was unyielding in her quest to understand her children's condition, and she made sure they were front and center in Folling's research priorities (Centerwall and Centerwall 2000 and Koch 2007).

In 1951, Horst Bickel, a professor at the Children's Hospital in Birmingham, England, diagnosed PKU in a toddler presenting with severe mental retardation. Again it was a mother advocating for her child that inspired the researcher; according to Professor Bickel's accounts, the child's mother continuously hounded him about treatment options (Bickel 1996). Bickel went on to hypothesize a causal relationship between excess levels of phenylalanine and the brain damage seen in PKU patients. These two mothers were perhaps the first "advocates" in newborn screening. They pushed the researchers and physicians to dig deeper to find the cause of—and ultimately a prevention strategy for—mental retardation in infants with PKU.

Moving past disease identification in the discovery phase, Bickel and his colleagues developed the first phenylalanine-free infant formula. Within a couple of months of starting the new regimen, physicians noted significant improvement in the cognitive function of the first toddler to receive the new formula (Bickel 1996). Once those who specialized in metabolic disorders recognized the success of the PKU formula, other clinicians and researchers acknowledged that early identification and subsequent administration of this intervention was effective in preventing negative sequelae. The question remained, however: how to accomplish early identification of infants who could benefit from the new formula? Advocates and researchers began to focus on developing screening methodologies.

The first screening tool developed was the "diaper test." In 1957, Dr. Willard Centerwall, the father of a child with mental retardation, discovered that if a solution of ferric chloride was applied to the wet diaper of a baby with PKU, it would produce a green color (Koch 1997). While this test provided a way to identify babies with PKU, it did not translate into a mass screening effort. Because the buildup of phenylalanine in urine often takes a few days after birth, many cases of PKU were missed at birth even when the diaper test was employed and were identified only after severe mental retardation had occurred.

In 1958, Dr. Robert Guthrie, considered the father of newborn screening, discovered that the bacterial-inhibition assay he had been using in cancer research could be modified to identify the presence of phenylalanine in the blood (Guthrie 1996). This test, which became known as the "Guthrie Spot," uses a very small amount of blood to determine the level of phenylalanine in the blood. This scientific innovation was slow to take hold in the broader medical community, perhaps because physicians were not aware of the test, or because they questioned the clinical validity and utility of the test, or considered the disease too rare to warrant screening, or did not trust the dietary intervention (Koch 1997). Yet within five years of Guthrie's discovery, PKU screening transitioned from limited use in the clinical setting to a state-mandated public health screening program for all infants born in Massachusetts. This initial milestone in newborn screening had as much to do with effective advocacy as it did the science and medical advancements that made screening for, and treating, PKU possible.

Dr. Guthrie was the father of a child with mental retardation (unrelated to PKU) and was quite active in the Buffalo Chapter of the New York State Association for Retarded Children (Guthrie 1996). Soon after his discovery, Guthrie's niece was diagnosed with PKU at 15 months of age. She developed severe cognitive impairment because her PKU was not identified at birth (Guthrie's assay had not been broadly implemented yet) (Koch 1997). This personal experience of the consequences of late diagnosis prompted Guthrie to take action. In 1961, he pitched the idea of a state-mandated screening program to the Director of Maternal and Child Health for New York State, who liked the idea and encouraged Guthrie to do a pilot project. By 1963, the project had expanded to 29 states; 400,000 infants had been screened, and 39 cases of PKU had been identified (Guthrie 1996).

Dr. Guthrie refused to profit from his invention, which helped to keep the cost low and made it easier to justify screening on a large scale (Paul 2008; Koch 1997). He would end up dedicating much of his life's work to making sure children had access to screening (Koch 1997). As these stories illustrate, the history of PKU research, screening, and treatment owes much to physicians and researchers whose lives had been touched by children affected by mental retardation. The transition of PKU testing from a voluntary private activity to a mandated public process underscores the important interaction between social and political advocacy.

THE SOCIAL AND POLITICAL CONTEXT

Innovation in science and technology does not occur in a vacuum; rather, it transpires within the context of political and social forces. Recognizing the multitude of factors involved in policy making—including advocacy—is essential to understanding the evolution of policy over time. As historian M. S. Lindee (2000, p. 237) notes, "Disease is often a place where a culture's moral narratives, social organization and economic pressures are made manifest. Just as a mutant fly provides a window into the genome, so disease provides a window into culture."

Guthrie was promoting the idea of state-mandated newborn screening programs at a time when sociocultural forces were in his favor. By the late 1950s, there was a push to treat mental illness in a community setting rather than in an institution. This represented a distinct shift from previous approaches to psychiatric illness (Grob 1987). By the early 1960s, President John F. Kennedy, who had an institutionalized cognitively impaired sister, was on record as supporting increased federal involvement in mental health policy (Paul, 2008; Grob, 1987). Not only was President Kennedy sympathetic to families dealing with this issue, but mental retardation was a topic that weighed heavily on the minds of many. House of Representatives leadership was supportive of research on the biological and genetic factors behind mental retardation; by 1970, in an effort to find new approaches to treating people with mental illness and mental retardation, federal grants were issued to develop new community programs and construct more than 500 mental health facilities around the country (Grob 1987; Paul 2008).

The Health Resources and Services Administration funded Dr. Guthrie's early work (Therrell Jr. 2001), as did the National Association for Retarded Children, later known as the National Association for Retarded Citizens (NARC) (Koch 1997). NARC encouraged state leaders to rally around the idea of mandated PKU screening, and was viewed as a crucial ally to advocates who were working to expand PKU screening (Guthrie 1996). Guthrie and others traveled state to state to talk with public health leaders about mandating PKU screening. Backed by NARC and the March of Dimes, and together with other advocates, consumers, and industry leaders, Guthrie was able to apply pressure on state lawmakers. State-mandated PKU screening programs began to emerge across the country, first in Massachusetts and then in New York. By the late 1970s, PKU screening was mandated in all 50 states (Howse, Weiss, and Green 2006). The high profile of NARC combined with presidential support provided political capital, and the involvement of Guthrie and other advocates provided the much needed social capital.

Newborn Screening, Beyond PKU

As the effort to mandate newborn screening state-by-state continued, Dr. Guthrie adapted his PKU test and identified new ways of screening for more than 30 other conditions (Koch 1997). Still, state legislatures and medical and public health communities were hesitant to mandate screening for additional conditions. Prior to 1968, when the World Health Organization (WHO) published *Principles and Practice of Screening for Disease*, there were no accepted guidelines for determining scientific validity in the context of large-scale, publicly funded screening programs (Wilson and Junger 1968). The WHO paper laid out the aim of early disease detection and described different screening methods and criteria for test evaluation and, most importantly, established a set of guidelines under which additions to newborn screening panels should be considered. These included principles such as: the condition should be an important health problem, there should be a reliable way to diagnose the condition and an acceptable treatment, and there should be a recognizable latent or early symptomatic stage (Wilson and Junger 1968). The authors highlighted the importance of these principles when screening is carried out by a public health agency. The WHO report received mixed reviews from the public health community: while some considered the principles a useful tool, others viewed the report as a roadblock to the expansion of newborn screening.

In 1975, the National Academy of Sciences (NAS) raised concerns regarding the potential risks of newborn screening in its publication *Genetic Screening: Programs, Principles and Research* (NAS 1975). There is some overlap between the 1968 principles and the 1975 NAS criteria: both require appropriate testing methods, both include availability of treatment as an important factor, and both address the cost-benefit aspect. One issue this set of criteria put forth that is important in the context of law and ethics is informed consent. The NAS expressed concerned about the role of public health and how this affects personal autonomy under a state mandated program, both autonomy of the child and parents' autonomy to determine what happens to their child.

Guthrie and his colleagues were steadfast in their support of expansion and recognized by 1980 that they were just getting started when it came to newborn screening (Bickel, Guthrie, and Hammersen 1980). Even though NAS and other expert organizations were hesitant, there was significant support for expanded newborn screening. Expansion continued throughout the 1980s to include conditions such as galactosemia, maple syrup disease, homocysteinuria, sickle cell disease, and cystic fibrosis. Today, roughly 50 years after the first state-mandated newborn screening began, state programs mandate screening for between 22 and 49 conditions; the majority of states (38) screen for 29 or more conditions (NNSGRC 2010).

The Role of Technology

Technological capabilities often develop much faster than the bioethics or public health community can respond. A prime example of this is the application of tandem mass spectrometry (MS/MS) to newborn screening in 1998. This new application of an existing technology provided laboratories with the ability to screen for a multitude of conditions using a single punch from a newborn screening blood specimen.

One of the catalysts for implementing MS/MS as a screening platform was the need for a testing platform capable of identifying infants at risk for medium-chain acyl-coenzyme A dehydrogenase (MCAD). Under fasting or stressful conditions, individuals with MCAD can slip into a coma, have seizures, suffer brain injury, or even die (Blois et al. 2005; Kahler and Fahey 2003). MCAD falls within the scope of newborn screening as traditionally defined because (1) it poses a serious risk of harm to the infant; (2) an effective intervention strategy exists; and (3) with the application of MS/MS, it can be detected using an analytically valid screening mechanism.

MS/MS machines vary in cost, but the cost is estimated to be around $250,000. There is also an initial investment in software and data management systems as well as the cost of annual maintenance. To use MS/MS in newborn screening programs, states require two to eight machines, depending on the volume of specimens being processed each day. Because the machines can process blood samples at a high capacity and can run multiple analyses on a single sample, the implementation of MS/MS technology opened the door for expanding screening for a host of other rare conditions. In the past, the ability to screen was limited to what could be identified using individual screening platforms for each condition; factors such as the amount of blood needed for testing, the labor required to process samples and tests, and the time involved in following up with families tended to restrict the scope of newborn screening. While the application of MS/MS resolved some of these issues, it also created new challenges, such as increased workloads for follow-up staff because of the increase in abnormal screening results.

This new high-throughput screening platform prompted the largest expansion of newborn screening to date. The concern over false positives (in which an infant has an abnormal screen and repeat or confirmatory testing later confirms the infant does not have the condition) and the clinical utility of screening results for some of the newly proposed conditions prompted many questions: Was enough known about these rare conditions to warrant screening all newborns? How would screening affect infants and families? What would be the long-term health benefits and potential risks associated with screening and treatment? What would be the impact of expanded screening programs on the public health and health care systems? The majority of conditions for which MS/MS is capable of screening had not been part of the newborn screening program prior to 1998 (Howse, Weiss, and Green 2006). There was uncertainty within the newborn screening community about how to deal with the growing range of testing possibilities. Should public health departments screen for all conditions for which screening is technically feasible? Or should they screen for conditions only when definitive treatment is available to change the course of the condition?

To obtain expert review of these questions, the Health Services Research Administration, the federal agency with primary responsibility for improving access to health care for vulnerable populations, contracted with the American College of Medical Genetics (ACMG). The resulting report, *Newborn Screening: Toward a Uniform Screening Panel and System*, was published in 2006. It recommended that 29 conditions (of the 84 that were evaluated) be included in all state-mandated newborn screening programs (ACMG 2006). In addition, the report identified an additional 25 conditions that states *should* consider adding to their screening panels.

The March of Dimes had an active presence during the development of the ACMG report and strongly supported its findings (Howse 2006). Yet, as was the case with previous reports on newborn screening, the ACMG report was met with some resistance from the academic and scientific communities. Concerns revolved around whether the 29 conditions for which screening would be mandated met the criteria previously set forth (Wilson and Junger 1968; NAS 1975 American Academy of Pediatrics 2000). In some cases, concern was over clinical utility and availability of sufficient evidence about treatment options. Other concerns included unclear clinical validity of the tests, limited knowledge of long-term outcomes, and an increase in false positive rates, which could negatively affect families.

The release of the ACMG report had a profound impact on states' newborn screening panels. By 2009, the majority of states had adopted most or all of the 29 conditions recommended for inclusion. In addition, the report marked another important shift for newborn screening advocates. Prior to the ACMG recommendations, advocates had to make the case for expanded newborn screening to each state's health department and/or state legislatures. The broad recommendation of the report to establish a national standard for newborn screening opened the door for advocates to target their efforts at the national level. As the science and policies evolve and new opportunities arise, so do the strategies of the advocates.

Who Is at the Table?

Newborn screening is not an issue of concern for the average citizen. As a result, the general public has not weighed in on the issue of expanding newborn screening programs. Although the goal of newborn screening is to screen every infant born in the United States, few parents are aware that such screening occurs. Davis et al. (2006) conducted 22 focus groups with 51 parents between 2003 and 2004 and found that parents "demonstrated little knowledge about newborn screening," and if they received information about newborn screening it was generally with a multitude of brochures in a packet being sent home with the parents. If newborn screening is not on the minds of new parents—the most likely group of people to be concerned about this topic—it is not a stretch to say it is most likely not forefront in the minds of people who are not parents or who perhaps have not been through childbirth recently. So who, then, is weighing in on newborn screening issues? The target audience is, of course, too young to communicate and the parents are largely unaware.

The responsibility of ensuring that the benefits of newborn screening outweigh the harms has fallen to clinical experts, researchers, and public health officials. Indeed, as one expert has commented, "[we] rely on public health experts to see the larger picture and take into account needs that may not be voiced by organized stakeholders, including our common need to avoid adopting valueless or even harmful products or interventions" (Paul 2008, pg 13). Public health departments are then responsible for securing a place at the table for representatives of those individuals and groups who may be affected by such decisions. An important voice in this process is the population the program serves—in this case, infants and families of the infants screened. Over time, states have welcomed parents into the decision-making process, and at present most of the state newborn screening committees include at least one or two parents of affected children.

For many rare diseases, the parents of affected infants are organized and impassioned, and their voices are well represented. This is a testament to the hard work of advocacy groups such as the Genetic Alliance, Cystic Fibrosis Foundation, Hunter's Hope Foundation, and March of Dimes. The March of Dimes has actively favored expansion, endorsing ACMG's call for national newborn screening guidelines. It has been the most vocal advocacy organization involved in supporting expansion of newborn screening (MOD, 2007).

A large scale screening program such as newborn screening will ultimately result in some false positive results. The number of false positives depends on the proportion of affected infants in the screened population and the nature of the test, but always increases as the number of screened infants increases. In order to assure that all affected infants are identified early to avoid negative health consequences, there will necessarily be infants who are falsely identified by the screening process as having one of the conditions. The impact of false positives on the families of these infants has received only modest attention, and their voices are missing from the vast majority of newborn screening dialogue.

In contrast with the advocacy community, which generally favors the expansion of screening, some in the scientific community have expressed concern that the expansion of newborn screening is moving too quickly. Other scholars argue that as new technology becomes available, we should seek to use it to its fullest potential. From this viewpoint, any expansion of screening programs is positive, and information that can alleviate a diagnostic odyssey is a benefit in and of itself (Howell 2006) and information provided by expanded newborn screening, even in the absence of clear treatment options, might serve as a useful benefit to parents (for more on the provision of genetic information as a benefit, see Chapter 6). Furthermore, identification of cases of the disorders may enable discovery and scientific development related to future treatment possibilities.

On the other hand, scholars argue that it is irresponsible to add conditions to the newborn screening panel without sufficient data to support the claim of public health benefit (Botkin et al. 2006). They contend that premature expansion of screening to include conditions for which effective treatment is not yet available will dilute the overall impact of screening programs, and may even cause harm. This cautionary view is represented by those who urge the newborn screening community to take the time to carefully weigh the features of each condition before mandating screening. Bioethicists and other academic researchers are concerned with the ethical, legal, and social implications of moving forward too quickly with genetic research and the application of new technologies. As a compromise solution, Botkin and colleagues (2006) propose that newborn screening expansion be done within a research paradigm to encourage the use of promising but unproven approaches to screening, but not at the expense of "opening programs to the indefinite use of ineffective or harmful technologies." This is where long-term health outcomes research can play an important role in newborn screening.

Given that scientific knowledge and technological advancements are creating a space into which newborn screening can further expand, the potential ethical, legal, economic and social implications are of even greater import. Some view such expansion as a benefit to society, while others see it as a potential harm. Technological advancements, such as MS/MS, have certainly fueled the debate among states over when to expand and for which conditions to screen. Future technological advancements that allow for faster, cheaper, and perhaps more accurate screening platforms will fuel even further expansion. And, since the debate has been pushing newborn screening as a national issue, some argue that newborn screening should be addressed as an issue of concern on the national agenda and not solely localized at the state level.

ADVOCACY, SCREENING, AND HEALTH DISPARITIES

Not all advocates carry the same weight or power. The stronger and more powerful voices often influence decision-makers. Sustained advocacy, particularly at the federal level, requires substantial resources. These resources come from parties interested in or affected by the cause of the advocate. When there are disparities in the resources available to different advocacy groups, these disparities translate into differential political influence and, therefore, political success. In particular, such differences often translate into increased disparities for groups with low socioeconomic status (SES). Pediatrician and policy analyst Lauren A. Smith documents how this self-reinforcing mechanism works in the discovery phase of research, noting that the dramatically low level of investment in sickle cell research relative to research on cystic fibrosis reflects at least in part the different resources available to the subgroups of the population affected by the two diseases—namely, African American populations and European American populations (Smith et al. 2006). As more funding flows into research for cystic fibrosis relative to sickle cell disease, there is an opportunity for increased health disparities between European Americans and African Americans.

When the word "disparity" is used in most settings, it is generally in relation to race/ethnicity, age, gender, or SES. Given that all infants born in the United States are screened under the auspices of public health programs regardless of race/ethnicity or SES, there is not much room for disparity in who gets screened. However, once an infant has screened positive, diagnostic testing is needed to confirm disease status. If the screen is determined to be a true positive, then the infant will require treatment or an intervention of some type. Interventions can range from an alteration in the infant's diet to invasive surgery. At this point, infants are subject to the same types of disparity that we know exist across the spectrum of health care services.

Third-party payers, both public and private, are responsible for covering newborn screening fees and follow-up care. In many states, close to half of all births are paid for by Medicaid, which means that taxpayers ultimately pay for the screening and diagnostic follow-up for infants with abnormal results. While some state newborn screening programs are completely self-funded by fees, other states rely on some general fund dollars to cover program costs. As with all public health programs, there are opportunity costs of using limited health care and public health dollars on any one particular service. In one sense, newborn screening provides a benefit to all newborns because screening determines whether or not an infant is at risk for one of these rare conditions. Knowing the infant is not at risk for one of these conditions has its own value. But ultimately, a small number of individuals experience a health benefit, which is simply a function of the rarity of the conditions identified.

Once an infant has been screened and diagnosed, its parent(s) are left to navigate the health care system. The ability of parents to effectively navigate the system may vary by SES, race/ethnicity, and geographic location (e.g., rural communities typically have less access to genetic services than city dwellers). Advocates for expansion have mainly focused on the number of conditions states include on newborn screening panels, but states' capacities to follow up on abnormal newborn screening results and connect affected infants with providers is an important determinant of the program's impact on infant health and health disparities. Some states offer financial assistance and support through genetic clinics, but the amount and type of available aid varies across the country. Without adequate infrastructure and funding to support both short- and long-term follow-up, states have an unequal ability to ensure that all infants receive appropriate treatment once identified.

WHAT ABOUT OUTCOMES RESEARCH?

The goals of most advocacy efforts for newborn screening have been to bring awareness to rare conditions that might otherwise be ignored, to encourage researchers and clinicians to develop methods for finding and diagnosing infants with these conditions, and to motivate public health officials and the medical community to design interventions to improve the health outcomes of affected infants. It is more difficult to find evidence of advocates insisting that newborn screening programs incorporate long-term health *outcomes* research. A few states have implemented long-term follow-up programs to track the progress of affected individuals overtime, however this does not necessarily translate into research on long-term health outcomes. Hoff (2008) found that "approximately 45% of state programs (17/38) reported conducting no activities past the point of confirming diagnosis . . ." (pg 3). For the minority of states that do conduct long-term follow-up, the activities vary by state and could include quality assurance, surveillance, or monitoring of care (Hoff 2008). Botkin and colleagues (2006) recommend that treatment protocols be coordinated and evaluated on a regional or national basis. This would allow for a more organized effort to analyze both the short- and long-term impact of newborn screening on the health of affected infants.

Given limited resources, advocates cannot focus on every aspect of the translational pathway. Still, the lack of advocacy related to outcomes research should be cause for reflection. In some cases, vocal and powerful advocates for disease identification may ignore or overlook significant data related to outcomes. The recent national debate over mammography recommendations provides an example of the difficulties associated with pitting scientific caution and risk/benefit analysis against the passion of disease advocates. Susan G. Komen for the Cure, an advocacy group that was founded to improve breast cancer awareness, has been very successful in generating increased attention to breast cancer as a leading killer of women. Its activities contributed to a fourfold increase in federal funding for breast cancer research in the 1990s (Braun 2003). Braun describes the impact of advocacy efforts on identifying breast cancer as a research priority, noting that "the third step, political action, became possible when breast cancer advocates joined together in the 1980s and 1990s to work toward legislative, regulatory, and funding changes, such as passage of the Mammography Standards Act and increased funding for the National Cancer Institute" (Braun 2003). In 2009, when the U.S. Preventive Services Task Force (USPSTF) revised its screening guidelines to recommend that women begin mammography at age 50, rather than 40, the change was met with significant backlash. Scientists and researchers had gathered data and developed sophisticated methods of combining large datasets and utilizing meta-analyses. The statistical evidence demonstrated that while a small percentage of women in their 40s benefit from early detection of cancer through mammography screening, a larger percentage of this age group actually experiences some type of harm, through over diagnosis, unnecessary biopsies, and/or chemotherapy and other cancer treatments that may not have been needed (Kolata 2009). But the USPSTF recommendation was viewed by breast cancer advocates as a step backward, and one that would likely reduce women's access to what they perceived as a valuable preventive care intervention.

Outcomes data are of interest to scientists and public health practitioners, but not only do they not attract passionate advocates, they are often viewed as a threat. This is also the case in relation to newborn screening. The voices of advocates pushing for expansion of newborn screening are strong, impassioned, and organized. Their efforts can sometimes overpower the cautionary voices—individuals and groups concerned about false-positive rates, unnecessary treatment, and/or family stigma.

CONCLUSION

Advocacy is a powerful force in shaping public policy. Advocates can be affected individuals and their families, organized advocacy groups, researchers, scientists, clinicians, and industry. Advocates embody power as a function of their organization, their communication strategies, and their resources, as well as the strength of their message. Voices that lack these assets are often not heard. Further, as we have shown, advocates are often more interested in disease identification and the provision of technology or treatment than in outcomes research about the long-term effectiveness of the technology or treatment.

The history of newborn screening policy underscores the powerful role of experienced and persistent advocates. Highlighting the historical role advocacy has played in shaping newborn screening policy provides insight for researchers and public health practitioners on how to work effectively with advocates and advocacy organizations, as well as what can happen when advocacy is absent. The scientific community can benefit from a better understanding of and appreciation for who is at the table and who is missing from the decision-making process. Understanding the role of science in relation to other political forces in shaping policy will benefit newborn screening as well as other types of genetic services.

REFERENCES

American Academy of Pediatrics. (2000). Serving the family from birth to the medical home. A report from the Newborn Screening Task Force convened in Washington DC, May 10-11, 1999. *Pediatrics.* 106(2 Pt 2):383–427.

[ACMG] American College of Medical Genetics. (2006). Newborn screening: toward a uniform screening panel and system. *Genet Med.* 8(Suppl 1):1S–252S.

Bickel H. (1996). The first treatment of phenylketonuria. *Eur J Pediatrics.* 155(Suppl 1): S2–S3.

Bickel H, Guthrie R, Hammersen G, eds. (1980). *Neonatal Screening for Inborn Errors of Metabolism.* Berlin Heidelberg New York: Springer-Verlag.

Blois B, Riddell C, Dooley K, Dyack S. (2005). Newborns with C8-acylcarnitine level over the 90th centile have an increased frequency of the common MCAD 985A>G mutation. *J Inherit Metab Dis.* 28(4):551–556.

Botkin JR, Clayton EW, Fost NC, et al. (2006). Newborn screening technology: proceed with caution. *Pediatrics.* 117(5):1793–1799.

Braun S. (2003). The History of Breast Cancer Advocacy. *Breast J.* 9(Suppl 2): S101–S103.

Centerwall S, Centerwall R. (2000). The discovery of phenylketonuria: the story of a young couple, two retarded children, and a scientist. *Pediatrics.* 105(1 Pt 1):89–103.

Davis T, Humiston SG, Arnold CL, et al. (2006). Recommendations for effective newborn screening communication: results of focus groups with parents, providers, and experts. *Pediatrics*. 117(5 Pt 2):S326–S340.

Grob GN. (1987). The forging of mental health policy in America: World War II to new frontier. *J Hist Med Allied Sci*. 42(4):410–446.

Guthrie R. (1989). Techniques and efficacy of screening: newborn screening. *Pediatrics*. 83(5):836–838.

Guthrie R. (1996). The introduction of newborn screening for phenylketonuria: a personal history. *Eur J Pediatrics*. 155(Suppl 1):S4–S5.

Hoff, T. (2008). "Long-term follow-up culture in state newborn screening programs," Genetics in Medicine, 10(6):1–9.

Howell R. (2006). We need expanded newborn screening. *Pediatrics*. 117:1800–1805.

Howse JL, Weiss M, Green NS. (2006). Critical role of the March of Dimes in the expansions of newborn screening. *Ment Retard Dev Disabil Res Rev*. 12(4):280–287.

Kahler SG, Fahey M. (2003). Metabolic disorders and mental retardation. *Am J Med Genet C Semin Med Genet*. 117(1):31–41.

Koch J. (1997). *Roberth Guthrie, the PKU story, a crusade against mental retardation*. Pasadena, CA: Hope Publishing House.

Kolata G. (2009). Behind cancer guidelines, quest for data. *New York Times*. November 22: A19. http://www.nytimes.com/2009/11/23/health/23cancer.html?pagewanted=2&_r=1. Accessed July 14, 2010.

Lindee MS. (2000). Genetic disease since 1945. *Nat Rev Genet*. 1(3):236–241.

March of Dimes (MOD) (2007). "Nearly 90 percent of babies receive recommended newborn screening tests," http://www.marchofdimes.com/aboutus/22663_25778.asp. Accessed December 11, 2010.

National Academy of Sciences (NAS). Genetic screening: programs, principles, and research, Washington, DC: 1975.

[NNSGRC] National Newborn Screening and Genetics Resource Center. NNSGRC Web site. http://genes-r-us.uthscsa.edu/. Accessed December 28, 2010.

Paul DB. (2008). Patient advocacy in newborn screening: continuities and discontinuities. *Am J Med Genet C Semin Med Genet*. 148C:8–14.

Scriver CR. (2001). Garrod's foresight; our hindsight. *J Inherit Metab Dis*. 24:93–116.

Smith LA, Oyeku SO, Homer C, Zuckerman B. (2006). Sickle cell disease: a question of equity and quality. *Pediatrics*. 117(5):1763–1770.

Therrell BL Jr.,(2001). U.S. newborn screening policy dilemmas for the twenty-first century. *Mol Genet Metab*. 74(1–2):64–74.

Wilson JMG, Junger G. (1968). *Principles and Practice of Screening for Disease*. Public Health Papers No. 34. Geneva: World Health Organization.

10

What Outcomes? Whose Benefits?

WYLIE BURKE AND NANCY PRESS

The ultimate goal of translational research is to protect and improve the health of the population (Zerhouni 2005). To fully accomplish this goal, research must address the striking disparities in health that occur both globally and within developed countries (e.g., Murray et al. 2006). Around the world, disease burden and life expectancy vary substantially by socioeconomic status, gender, place of residence, and race (Dressler, Oths, and Gravlee 2005; Haley 2006; Kilbourne et al. 2006; Lurie and Dubowitz 2007). These disparities are a particular concern at the end of the translational process, where research focuses on documenting and describing health outcomes and on evaluating the effectiveness of efforts to improve health. Do the new tests, drugs, or procedures that have been produced by translational research actually make people healthier? And, in the context of health disparities, do these innovations reach all who need them?

Health disparities research has typically been carried out by experts in public health and the social sciences, but more recently genome scientists have also begun to identify a role for genomic research in addressing these inequalities (Adeyemo and Rotimi 2010; Collins et al. 2003; Risch 2006). This chapter considers what a genomic approach to health disparities might look like, and what efforts would be needed to make such investigations successful.

THE CONTESTED ROLE OF GENOMICS IN HEALTH DISPARITIES RESEARCH

Many factors contributing to health disparities have been identified, including differential access to employment, education, and quality health care; racial discrimination; and differential exposure to environmental hazards (Dressler, Oths, and Gravlee 2005). There is also increasing evidence that health is correlated with social position: on average, wealthier and better-educated people are also healthier (Banks et al. 2006; WHO 2008). The primary focus of efforts to reduce disparities has therefore been on policy solutions and public health interventions, such as improved funding for health care, improved air quality, and community-based programs to improve health care and motivate healthy lifestyles (Banks et al. 2006; Lavizzo-Mourey and Knickman 2003; Lurie and Dubowitz 2007). A recent report from the World Health Organization also calls for systematic efforts to improve daily living conditions and address economic and power inequities (WHO 2008).

Following another paradigm, however, genome scientists suggest that genetic differences may contribute to health disparities, and that the relative importance of genetic factors versus social or environmental influences is unknown for most diseases (Burchard et al. 2003). This view comports with the fact that, in the United States and other developed countries, ethnicity and race are strongly associated with health disparity, and genetic differences between racial and ethnic groups are readily found. Gene variants associated with a wide variety of common diseases, including coronary artery disease, some cancers, multiple sclerosis, and diabetes, show marked differences in frequency across different racial and ethnic populations (e.g., Risch 2006; Adeyemo and Rotimi 2010). These findings have led genome scientists to promote research on group difference; for example, Risch et al. (2002, p.11) wrote, "Both for genetic and non-genetic reasons, we believe that racial and ethnic groups should not be assumed to be equivalent, either in terms of disease risk or drug response. A 'race-neutral' or 'color-blind' approach to biomedical research is neither equitable nor advantageous, and would not lead to a reduction of disparities in disease risk or treatment efficacy between groups".

Health disparities researchers working in the social sciences and public health are skeptical of a genomic solution to inequalities in health status (Krieger 2005). Genetic differences, they argue, cannot be disentangled from social correlates of race. For example, individuals from minority populations are disproportionately exposed to a wide range of social conditions associated with poor health status, including poor maternal and early childhood nutrition, poor quality education and health care, low income, and environmental hazards (WHO 2008). Therefore, even when genetic differences can be defined, their role in creating health disparities is open to question. This caution is particularly relevant with respect to common diseases such as diabetes, cancer, and heart disease. Although many gene variants associated with these diseases have been found, they typically explain only a small proportion of disease risk (Manolio et al. 2009).

Health disparities researchers are also concerned about the potential harms that could result from genomic approaches. If gene variants are used primarily to "explain" differences in health outcomes, genomic research is unlikely to reduce disparities; on the contrary, such research could be used to define health disparities as an expected and inherent aspect of population differences. An Australian public health researcher posits, for example, that genetic investigations have led to "a common view among doctors and nurses that diabetes in Aborigines is 'genetic' and therefore inevitable," with the result that effective public health measures for diabetes prevention have not been pursued in the Aboriginal population (McDermott 1998, p.1190). To the extent that genetic research has this effect, it will become the latest in a long (and discredited) line of scientific findings invoked to explain minority disadvantage as the result of inherent biological differences (Krieger 2005). Even if genetics were not used as an excuse for accepting or ignoring health disparities, however, placing a priority on genomic investigation might nonetheless direct resources away from other, arguably more promising, research on social determinants of health.

For all these reasons, researchers engaged in addressing health disparities appear to see little value in genomic approaches to the problem. But are they turning a blind eye to the benefits? Many experts anticipate that genomics will yield major advances in disease prevention. For example, as Deverka and Veenstra show in Chapter 5, genetic tests are already available that can improve drug treatment or identify individuals who would benefit from tailored prevention efforts, and many more such tests are expected (Guttmacher and Collins 2005; Peltonen and McKusick 2001; Roden et al. 2006). Equally important, genomic research provides new methods for understanding disease biology, and thereby plays an increasingly central role in biomedical research (Zerhouni 2005). Many genome scientists argue, for example, that genomics offers unique opportunities to understand the interplay between genetics and environmental risk (Adeyemo and Rotimi 2010; Bamshad 2005; Collins et al. 2003). Is it wise to assume that this powerful line of research has nothing to offer for health disparities?

We suggest that the key to a productive role for genomics in health disparities research is to shift from an emphasis on genetic causation to one that focuses on improving health outcomes. The central research question would become, how could genomic research add value to current efforts to reduce health disparities? In this approach, studies of genomic contributors to disease would be integrated into broader inquiries about the social determinants of health, rather than viewed as an alternative to such research. To understand how this shift might change translational genomic research, we need to consider the different pathways by which genome research can ultimately achieve health benefit.

Genetic tests represent the most rapid example of genomic translation (Burke, Laberge, and Press 2010; Khoury et al. 2009). Some tests identify genetic susceptibilities that be used to can guide drug therapy or preventive care. Genetic tests are also being developed for disease classification, allowing in some cases for more-precise treatment plans. Health care benefits from genetic testing are likely to increase over time. However, the potential benefits of genomic research extend well beyond the production of genetic tests. The use of genetic analysis to identify the biologic pathways involved in different disease states is providing new understanding of how diseases occur. This research has significant promise for identifying new avenues to drug development and disease prevention. From the perspective of health disparities, the most important contribution of genomics and associated molecular tools will be in their application to the study of the mechanisms by which environmental factors, including social determinants, affect health.

To take full advantage of these opportunities, genome scientists and health disparities researchers will need to engage in interdisciplinary work. A rigorous investigation of the interaction between social, environmental, and genetic determinants of health will require the development of collaborative projects that combine genome science with public health, medical, or social science methods (Gehlert et al. 2008; Hiatt and Breen 2008), and which attend to the perspectives of individuals and communities who experience disproportionate health burdens. These efforts could help to ensure that the experience and perspectives of populations experiencing health disparities are represented in study design and interpretation.

The dialogue required to achieve such collaborations will not be easy, because genomics and health disparities researchers have few points of contact and use different terminology, methods, and conceptual frameworks. But the potential payoff is large, if the engine of genomics can be harnessed to the task of reducing disparities. One of the early successes for genomic health research—progress in understanding the genetics of breast cancer—offers a window into the implications and challenges of this approach.

TESTING FOR GENETIC SUSCEPTIBILITY TO BREAST CANCER

Geneticists have long recognized families at high risk for breast cancer. They are characterized by certain distinctive clinical features: female family members experience a high rate of both ovarian cancer and early onset breast cancer, many affected women have bilateral breast cancer or breast and ovarian cancer, and male breast cancer may also be seen (Petrucelli et al. 1998). The discovery of the *BRCA1* and *BRCA2* genes in the early 1990s provided an explanation of these patterns: in most high-risk families, a deleterious mutation (or change) in *BRCA1* or *BRCA2* is inherited, causing increased breast and ovarian cancer rates. These mutations (hundreds have been identified in each gene) result in reduced function of the *BRCA1* or *BRCA2* protein, both of which play a role in maintaining the integrity of DNA. The reduced DNA-repair function is associated with cancer risk. Based on this knowledge, physicians now routinely offer *BRCA* testing to women with a strong family history of breast or ovarian cancer. If a mutation associated with cancer risk is found, it can inform choices about cancer care, screening, and the use of aggressive prevention measures such as surgical removal of ovaries or breasts.

However, minority women receive far less benefit from *BRCA* testing than white women do. One reason for the disparity is reduced access to testing: minority women experience a number of barriers to genetic testing related to their higher likelihood of living in poverty (Armstrong et al. 2005; Halbert, Kessler, and

Mitchell 2005; Kinney et al. 2006; Simon and Petrucelli 2009), including lack of access to information about testing, competing priorities, and cost (*BRCA* testing is about $3,500, and insurance coverage is variable). In addition, minority-serving physicians are less likely to offer testing (Shields, Burke, and Levy 2008), and the limited diversity among genetic counselors and medical geneticists may be a disincentive (Cooksey et al. 2005; Mittman and Downs 2008). In other words, barriers to *BRCA* testing are similar to barriers experienced by disadvantaged women to health care in general (Gerend and Pai 2008).

Even when minority women do obtain testing, they face an additional problem: they are more likely to receive an ambiguous test result, known as a "variant of unknown significance," or VUS (Easton et al. 2007; Haffty et al. 2006; Nanda et al. 2005; Weitzel et al. 2005). In the context of breast cancer testing, a VUS is a change in the *BRCA1* or *BRCA2* gene that cannot be clearly defined as either cancer-predisposing or a normal variant. This test result is clinically noninformative and frustrating for both patients and health care providers (Domchek and Weber 2008). Over time—often years—the testing laboratory may gain sufficient experience with the variant to be able to determine whether the woman tested is indeed at high risk or merely has a *BRCA* change that represents normal variation. But in the meantime, minority women are more likely to be left at the end of the testing process with information that is neither helpful nor reassuring, despite the high cost of the test. Because a lower proportion of minority women get meaningful test results, the cost of a useful result is overall more expensive.

An important reason for this discrepancy is that the research leading to *BRCA1/2* gene discovery was done predominantly in populations of European ancestry (see Chapter 3). Hundreds of different *BRCA* mutations have been identified (Petrucelli et al. 2007), and the limited studies in minority populations suggest that the mutation spectrum varies among different ancestral populations—so, we know a lot about *BRCA* and cancer risk in women of European descent, but not much about genetic risk for breast cancer in other groups. The lower rate of clinical testing among minority women adds to the problem: of the first 10,000 *BRCA* tests performed in the United States, less than 10% were for members of underrepresented minority populations (Frank et al. 2002). As a result of both these trends, the accumulation of knowledge about *BRCA* mutation spectrum and prevalence in minority populations has been slow. A vicious cycle ensues when low utilization of testing perpetuates limited knowledge about the *BRCA* mutation spectrum in minority populations, and thus perpetuates lower clinical value for the test. Arguably, then, *BRCA* testing has added to health disparities, rather than alleviating them (Hall and Olopade 2005).

However, *BRCA* testing is only a small part of the breast cancer story. *BRCA* mutations occur in only about 5% of women with breast cancer. Even if testing were as beneficial for minority women as for women of European ancestry, the effect on overall cancer outcomes would be small. Although increased inclusion of minority women in *BRCA* research is an important goal, a broader consideration of breast cancer disparities points to other, even more promising, opportunities for translational genomics.

BREAST CANCER IN MINORITY WOMEN

Epidemiologic studies of breast cancer reveal a paradox: the incidence of breast cancer is lower among disadvantaged women in the United States, whether disadvantage is assessed by socioeconomic status (SES) or by race and ethnicity (Harper et al. 2009)—yet poor and minority women are more likely to die when breast cancer is diagnosed (Clegg et al. 2009; Harper et al. 2009;).

Reduced quality of health care is part of the problem: mammography screening rates are lower among minority and low-SES women, especially those who do not have health insurance (Harper et al. 2009; Sabatino et al. 2008). Even when subsidies are available, mammography services may not reach all in need. A study of breast cancer screening programs in rural minority communities, for example, showed that inadequate road systems prevented access to mammography, both because mobile mammography units could not reach isolated communities and because travel to a health facility was often time consuming and burdensome for patients (R. James, personal communication, July 2009). As a result of such barriers, minority and low-SES women are more likely to be diagnosed with late-stage breast cancer (Clegg et al. 2009; Harper et al. 2009;). Treatment is also more likely to be substandard for poor or uninsured women (Freeman and Chu 2005). Therefore, much of the increase in breast cancer mortality among poor and low-SES women can be explained by lower screening rates or inadequate treatment; genomics cannot offer solutions to these problems.

However, a growing body of research points to different *types* of breast cancer as a factor in cancer disparities as well, and genomics may play a role in understanding their implications. In particular, minority and low-SES women appear to have an increased risk for breast cancers associated with early onset and high mortality—that is, cancers that are inherently more dangerous. Although genetic predisposition may play a role in the type of breast cancer a woman develops, several lines of research point to nongenetic factors as well. Animal studies document that social isolation and stress can increase cancer risk and affect tumor biology (Gehlert et al. 2008), and human studies also suggest a role for social deprivation in breast cancer outcomes. For example, studies show a higher rate of breast cancer types with poor prognosis in low-SES women irrespective of race (Parise et al. 2009; Vona-Davis and Rose 2009).

Methods to distinguish different breast cancer types are still evolving. The traditional approach is based on the presence of three cell receptors in the breast tumor: the receptors for the hormones estrogen and progestin, and for human epidermal growth factor receptor type 2 (HER2) (Dawson, Provenzano, and Caldas 2009). Breast cancers that lack all these receptors—so-called triple-negative tumors—tend to occur at a younger age than other breast cancers and are associated with higher rates of death. This kind of cancer is more common among women with mutations in the *BRCA1* gene. It is also more common among African American and Hispanic women, as well as women of low SES, in the absence of *BRCA1* mutations (Dawson, Provenzano, and Caldas 2009; Parise et al. 2009; Vona-Davis and Rose 2009).

A different way of assessing breast cancer type is based on *gene expression.* Most genes code for proteins—the building blocks that make up the cells, tissues, and organs of the body. The instructions for making the proteins are transmitted from the gene by messenger RNA (mRNA). Measuring mRNA levels is therefore a way of determining which genes are turned on, or expressed, in any given tissue. Applying this method to breast cancer, researchers have found differences in gene expression that can also distinguish breast cancer type (Dawson, Provenzano, and Caldas 2009). Basal-type breast cancer, one of the breast cancer types defined by this method, has been found more commonly in African American women and in women from Nigeria (Olopade et al. 2008). Like triple-negative breast cancer, basal cancer is associated with a higher mortality rate.

Triple-negative and basal breast cancer are partially overlapping categories. Most, but not all, triple-negative tumors have basal-type gene expression, and vice versa. It is likely that these two approaches to typing breast cancer are indicators of a more complex set of factors that collectively influence how rapidly and dangerously a breast cancer progresses (Dawson, Provenzano, and Caldas 2009). Why do triple-negative and basal breast cancer occur more commonly in disadvantaged women? Genetic susceptibility may be a factor: we know that women with *BRCA1* mutations are predisposed to triple-negative breast cancer; and there may be other inherited susceptibilities, perhaps more common in women of African and Hispanic descent, that also increase the risk for these deadly forms of breast cancer. Studies seeking to identify breast cancer susceptibilities unique to African American women are under way (Fejerman et al. 2009; Min et al. 2009)

However, the accumulating evidence on the effect of social factors such as isolation and stress on cancer risk and tumor biology (Gehlert et al. 2008), and of low SES on breast cancer type as well as on survival (Parise et al. 2009; Vona-Davis and Rose 2009), argue that conceptualizing genetics as separate from social/environmental factors—that is, as an either/or proposition—is erroneous. The greatest opportunities for genomic research may lie in finding the complex pathways by which factors that seem highly "social," rather than biological, act on the body in ways that produce biologic effects.

IMPORTANCE OF INTEGRATING SOCIAL AND BIOLOGICAL STUDIES

Freeman and Chu (2005) noted a number of social factors that contribute to cancer disparities, including factors related to poverty (such as difficulty accessing health care), culture (such as distrust of the health care system based on historical mistreatment), and social injustice (such as effects of racial prejudice). Gerend and Pai (2008) noted that each of the domains is multidimensional. In addition to creating financial barriers to care, for example, poverty is also associated with broader health systems effects: minority-serving physicians, for example, are less likely to order preventive care, and health care facilities may be located distant from the patient's work or home. People in poverty may also have pressing survival needs, such as obtaining food and shelter or maintaining hourly employment, that have a higher priority than health care. In addition to distrust, cultural factors that can affect health outcomes include common assumptions that tend to diminish health care participation, such as the belief among some African American women that they have a lower breast cancer risk than white women—a belief that is accurate for breast cancer incidence, but not for breast cancer mortality. Social injustice may play a role via institutionalized racism, or potentially via biological effects that stem from the stresses of racial discrimination or inferior social position (Banks et al. 2006; Gerend and Pai 2008). This last possibility—that social factors may produce adverse physiological effects, leading to an increased burden of disease—represents an important opportunity for molecular genetic investigation.

In the "downward causal" model of breast cancer risk proposed by researchers at the University of Chicago (Gehlert et al. 2008), social isolation, poverty, and other social factors, with their associated psychological states, are postulated to create physiological stresses that interact with individual predispositions to produce cancer. Disease outcomes are further affected by social factors that influence the timing of diagnosis and quality of treatment (Gehlert et al. 2008). In this model, genes, social structure, and environmental hazards all play a role in determining which individuals develop breast cancer, what type of cancer occurs, and what the outcome is.

POTENTIAL CONTRIBUTIONS OF GENOMICS

Although biological effects of social isolation, poverty, and other social determinants of health can be inferred from epidemiology, animal studies, and clinical observation (Banks et al. 2006; Gehlert et al. 2008; Vona-Davis and Rose 2009), much more information is needed to identify how social determinants affect biology—specifically, how they interact with other risk factors to produce higher disease rates—in order to identify the strategies that might be used to reverse them. It is within this context that the potential role of for genomics in health disparity research needs to be considered.

Genomics provides an increasing array of tools and research strategies that can aid in the investigation of models of multidimensional causality that recognize the powerful interactions between genetic and social determinants of health. For example, the study of correlations between social or environmental exposures and gene expression (what genes are turned on or off in different circumstances) may provide important insights into how social disadvantage is embodied in ways that affect health (Krieger 2005; Williams et al. 2010). In this respect, two innovative areas of genomic research may play a key role in understanding health disparities: the study of the human microbiome (Turnbaugh et al. 2007) and investigation into the disease implications of epigenetics (Esteller 2006).

The first of these, human microbiome studies, derives from the observation that microbial cells outnumber human cells by an estimated factor of 10 to 1 within the body of a healthy adult (Turnbaugh et al. 2007). It follows that differences in the *microbiome*—the microbial cells present in the mouth, gastrointestinal tract, skin, and other parts of the body—could have health consequences. The implications for health disparities are potentially significant, because factors such as diet, living conditions, and environmental exposures presumably influence the array of microorganisms in a given person's microbiome.

The second is the concept of epigenetic change Epigenetics refers to environmentally induced changes that have sustained effects on gene expression (hence the term *epi-genetic*, from the Greek *epi*, meaning over or above, the genetic). Some epigenetic phenomena are part of natural development. An example is gene methylation, a process in which a methyl group binds to DNA. Methylation suppresses gene expression and is part of normal embryological development, where it plays an important role in assuring the appropriate differention of the tissues of the body (lungs, heart, skin, etc.). Although DNA methylation and other changes in the structures surrounding genes serve a normal function, they may also go awry at times. Diet and other environmental conditions may affect whether genes are methylated, with potential adverse health effects. For example, some studies suggest that triple-negative breast cancer might result from the methylation of *BRCA1* or related genes (Dawson, Provenzano, and Caldas 2009). By silencing genes in the *BRCA1* pathway, methylation would mimic the presence of a *BRCA1* mutation, so that an environmental effect could produce the same cancer risk as an inherited mutation.

Importantly, epigenetic effects may be passed from parent to child, even though they do not alter the DNA sequence. Situated between the many facets of the environment—whether a person lives in a polluted city, whether his or her job involves physical activity or exposure to certain chemicals or other hazards, what kind of insulation is in the person's house—and the DNA sequence, epigenetics is the point at which nature and nurture intersect in the body. To the extent that adverse social environments produce epigenetic effects—as with methylation potentially leading to risk for triple-negative breast cancer—they could play a central role in health disparities. Importantly, epigenetic effects can be misinterpreted as inherited effects if not fully evaluated. Epigenetic research may play a crucial role in distinguishing genetic from environmental contributors to health disparities and, in the process, help to identify environmental risks for public health action.

PLACING GENOMICS IN CONTEXT

Better understanding of the physiological effects of the social environment will not necessarily reduce disparities if the underlying cause is social inequity. To contribute to solutions, genomic research must be integrated into a larger effort that considers the potential for benefit from different lines of research and policy development. From the perspective of health disparities, we might ask, what is the effect on different populations of pursuing one translation opportunity versus another? What would have to be in place for the application to be useful to underserved populations?

Answering these questions will inevitably be an iterative process. As Eisenberg (1999, p. 1868) noted, "Technology is rarely inherently good or bad, always or never useful. The challenge is to evaluate when in the course of illness it is effective, for whom it will enhance outcomes, and how it should be implemented or interpreted." This comment focuses the outcomes challenge for translational research. The most important contribution of genomics to the problem of triple-negative or basal breast cancer would be to identify actions or tools that could reduce the incidence and risk of these types of cancer—for example, the identification of specific environmental or social factors that could be addressed effectively by public health action, or new opportunities for the development of effective targeted therapy.

IMPLICATIONS FOR TRANSLATIONAL RESEARCH

Translational research is by its nature a directed process with a specific goal: the production of new tools to promote health. Using knowledge of outcomes to inform discovery research can help researchers clarify the intended benefits, including the need for tools that account for the realities of health disadvantage associated with inner-city and rural settings, racism and other social inequities, cultural diversity, and underfunded health systems. To succeed, this process will need to advocate both for appropriately focused translational research and for policies, institutions, activities, and systems—informed by research—that enable improvements in health.

In considering the contribution of genomics to this goal, it is helpful to consider the challenges at different phases of research. In the discovery phase, the most obvious concern is to ensure that genomic research includes participants from disadvantaged groups, as Chapter 3 argues. Achieving this objective will require both resources and partnership development (Hughes et al. 2004; James et al. 2008) To engage effectively with members of disadvantaged communities about the process of research and its potential benefits, genome science researchers will need to communicate effectively about the value of research and the rationale for participation from diverse populations.

To pursue research that is relevant to health disparities, researchers also need to develop a nuanced understanding of the health outcomes of interest and of contributing social and environmental factors. Mere information exchange is insufficient; to ensure that translational research meets its objectives, the dialogue between genome scientists, health disparities researchers, and communities will need to overcome barriers of language, power, and position. Individuals who can serve as trustworthy intermediaries are necessary to help researchers and vulnerable communities understand each other's viewpoints and identify common ground. The need for dialogue also extends beyond the initiation of the research process: measures for oversight of sample use and data sharing, endorsed by participants and their representatives, are also essential to successful research (Arbour and Cook 2006; Boyer et al. 2005).

Different challenges arise in the development stage of research. Key among these is how the needs of disadvantaged populations can be accommodated in commercial development of drugs and other health care products. By definition, these groups lack the size and financial resources that make for an attractive market. If health disparities are to be reduced, innovative approaches to product development will be needed—possibly private–public partnerships or incentives such as exist for orphan drugs—to support development directed toward innovations designed for resource-poor environments. A possible model is the effort of the USA-India Chamber of Commerce to create cross-border partnerships aimed at developing affordable drugs and other health interventions (USA-India Chamber of Commerce 2010).

Finally, the study of health outcomes underscores the importance of meaningful interchange between researchers working at different stages of the translational cycle. The study of population differences in health has the potential to identify new lines of inquiry for research at the discovery and development stages (Hiatt and Breen 2008). Exchanges are likely to be more rewarding to the extent that they incorporate the perspectives and preferences of communities experiencing health disparities. Indigenous knowledge may, for example, play a critical role in identifying the social changes or environmental hazards that are most important to explore in studies evaluating interactions between genes and environment, a point illustrated in Chapter 11.

Promoting dialogue across disciplinary divides is not easy. Finding common ground may be particularly challenging when different parties bring conflicting assumptions to the table—as when genomic researchers assume that studies of genetic cause provide the primary route to reducing health disparities, or when public health researchers assume that genomics has no role to play in this effort. Nor is it easy to overcome the social, cultural, and economic divide between universities and communities in order to incorporate the views of disadvantaged communities in developing research agendas. Without efforts to overcome these communication challenges, however, genomic research may continue to be largely isolated from the growing knowledge about social determinants of health (WHO 2008), and health disparities research may lack the benefits to be derived from a growing array of powerful molecular tools.

CONCLUSION

At all stages of translational research, communication strategies are needed to assist genomic researchers in identifying how genomic tools can contribute to addressing the complex problems at the heart of health disparities. Successful research strategies are likely to include several components: (1) studies of genetic risk that incorporate interactions with social and environmental factors; (2) the use of genomic and other molecular tools to define the biological effects of social factors, including effects on gene expression and the microbiome; and (3) cross-disciplinary and university–community partnerships.

Partnerships matter because the measurement of outcomes is a complex undertaking, requiring careful consideration of different stakeholder perspectives and health care contexts. An expensive drug or procedure that provides a small incremental benefit might be viewed favorably in a well-funded environment, or by the developer who stands to profit from its use, but the same intervention might create unacceptable opportunity costs in a resource-poor environment. Similarly, genetic risk assessment might provide value if the test is designed to detect specific actionable risks, but it might be more questionable if the assessment also produces ambiguous risk information that consumes clinician time and health care resources. Benefits and harms for genetic services are highly dependent on social as well as clinical context, and they may vary strikingly for different individuals or health care settings. The application of new knowledge about gene–environment interactions or epigenetic effects is likely to produce more benefit if it incorporates robust knowledge about social factors and about the preferences of disadvantaged populations. Achieving an active dialogue will be challenging, but it promises rich rewards in understanding and addressing health disparities.

REFERENCES

Adeyemo A, Rotimi C. (2010). Genetic variants associated with complex human diseases show wide variation across multiple populations. *Public Health Genom*. 13:72–79.

Arbour L, Cook D. (2006). DNA on loan: Issues to consider when carrying out genetic research with aboriginal families and communities. *Community Genet*. 9(3): 153–160.

Armstrong KE, Micco E, Carney A, Stopfer J, Putt M. (2005). Racial differences in the use of *BRCA1/2* testing among women with a family history of breast or ovarian cancer. *JAMA*. 2293(14):1729–1736.

Bamshad M. (2005). Genetic influences on health: does race matter? *JAMA*. 294(8): 937–946.

Banks J, Marmot M, Oldfield Z, Smith JP. (2006). Disease and disadvantage in the United States and in England. *JAMA*. 295(17):2037–2045.

Boyer BB, Mohatt GV, Lardon C, et al. (2005). Building a community-based participatory research center to investigate obesity and diabetes in Alaska Natives. *Int J Circumpolar Health*. 64(3):281–290.

Burchard EG, Ziv E, Coyle N, et al. (2003). The importance of race and ethnic background in biomedical research and clinical practice. *N Engl J Med*. 348(12): 1170–1175.

Burke W, Laberge AM, Press N. (2010). Debating clinical utility. *Public Health Genomics*. 13(4):215–223.

Clegg LX, Reichman ME, Miller BA, et al. (2009). Impact of socioeconomic status on cancer incidence and stage at diagnosis: selected findings from the surveillance, epidemiology, and end results: National Longitudinal Mortality Study. *Cancer Causes Control*. 20(4):417–435.

Collins FS, Green ED, Guttmacher AE, Guyer MS; U.S. National Human Genome Research Institute. (2003). A vision for the future of genomics research. *Nature.* 422(6934):835–847.

Cooksey JA, Forte G, Benkendorf J, Blitzer MG. (2005). The state of the medical geneticist workforce: findings of the 2003 survey of American Board of Medical Genetics certified geneticists. *Genet Med.* 7(6):439–443.

Dawson SJ, Provenzano E, Caldas C. (2009). Triple negative breast cancers: clinical and prognostic implications. *Eur J Cancer.* 45(Suppl 1):27–40.

Domchek S, Weber BL. (2008). Genetic variants of uncertain significance: flies in the ointment. *J Clin Oncol.* 26(1):16–17.

Dressler WW, Oths KS, Gravlee CC. (2005). Race and ethnicity in public health research: models to explain health disparities. *Annu Rev Anthropol.* 34:231–252.

Easton DF, Deffenbaugh AM, Pruss D, et al. (2007). A systematic genetic assessment of 1,433 sequence variants of unknown clinical significance in the *BRCA1* and *BRCA2* breast cancer-predisposition genes. *Am J Hum Genet.* 81(5):873–883.

Eisenberg J. (1999). Ten lessons for evidence-based technology assessment. *JAMA.* 282(19):1865–1869.

Esteller M. (2006) The necessity of a human epigenome project. *Carcinogenesis.* 27: 1121–1125.

Fejerman L, Haiman CA, Reich D, et al. (2009). An admixture scan in 1,484 African American women with breast cancer. *Cancer Epidemiol Biomarkers Prev.* 18(11): 3110–3117.

Frank TS, Deffenbaugh AM, Reid JE, et al. (2002). Clinical characteristics of individuals with germline mutations in *BRCA1* and *BRCA2*: analysis of 10,000 individuals. *J Clin Oncol.* 20(6):1480–1490.

Freeman HP, Chu KC. (2005). Determinants of cancer disparities: barriers to cancer screening, diagnosis, and treatment. *Surg Oncol Clin N Am.* 14(4):655–669.

Gehlert S, Sohmer D, Sacks T, Mininger C, McClintock M, Olopade O. (2008). Targeting health disparities: a model linking upstream determinants to downstream interventions. *Health Aff (Millwood).* 27(2):339–349.

Gerend MA, Pai M. (2008). Social determinants of black-white differences in breast cancer mortality. *Cancer Epidemiol Biomarkers Prev.* 17(11):2913–2923.

Guttmacher AE, Collins FS. (2005). Realizing the promise of genomics in biomedical research. *JAMA.* 294(11):1399–1402.

Haffty BG, Silber A, Matloff E, Chung J, Lannin D. (2006). Racial differences in the incidence of *BRCA1* and *BRCA2* mutations in a cohort of early onset breast cancer patients: African American compared to white women. *J Med Genet.* 43(2):133–137.

Halbert CH, Kessler LJ, Mitchell E. (2005). Genetic testing for inherited breast cancer risk in African Americans. *Cancer Invest.* 23(4):285.

Haley LL Jr. (2006). Stuck in neutral: continued challenges with healthcare disparities. *Acad Emerg Med.* 13(2):191–194.

Hall M, Olopade OI. (2005). Confronting genetic testing disparities: knowledge is power. *JAMA.* 293(14):1783–1785.

Harper S. Lynch J, Meersman SC, Breen N, Davis WW, Reichman MC. (2009). Trends in area-socioeconomic and race-ethnic disparities in breast cancer incidence, stage at diagnosis, screening, mortality, and survival among women ages 50 years and over (1987–2005). *Cancer Epidemiol Biomarkers Prev.* 18(1):121–131.

Hiatt RA, Breen N. (2008). The social determinants of cancer: a challenge for transdisciplinary science. *Am J Prev Med.* 35(2S):S141–S150.

Hughes C, Peterson SK, Ramirez A et al. (2004). Minority recruitment in hereditary breast cancer research. *Cancer Epidemiol Biomarkers Prev.* 13(7):1146–1155.

James RD, Yu JH, Henrikson NB, Bowen DJ, Fullerton SM; Health Disparities Working Group. (2008). Strategies and stakeholders: minority recruitment in cancer genetics research. *Community Genet.* 11(4):241–249.

Khoury MJ, McBride CM, Schully SD et al. (2009). The Scientific foundation for personal genomics: recommendations from a National Institutes of Health-Centers for Disease Control and Prevention multidisciplinary workshop. *Genet Med.* 11(8):559–567.

Kilbourne AM, Switzer G, Hyman K, Crowley-Matoka M, Fine MJ. (2006). Advancing health disparities research within the health care system: a conceptual framework. *Am J Public Health.* 96(12):2113–2121.

Kinney AY, Simonsen SE, Baty BJ, et al. (2006). Acceptance of genetic testing for hereditary breast ovarian cancer among study enrollees from an African American kindred. *Am J Med Genet A.* 140(8):813–826.

Krieger N. (2005). Stormy weather: race, gene expression, and the science of health disparities. *Am J Public Health.* 95(12):2155–2160.

Lavizzo-Mourey R, Knickman JR. (2003). Racial disparities—the need for research and action. *N Engl J Med.* 349(14):1379–1380.

Lurie N, Dubowitz T. (2007). Health disparities and access to health. *JAMA.* 297(10):1118–1121.

Manolio TA, Collins FS, Cox NJ, et al. (2009). Finding the missing heritability of complex diseases. *Nature.* 461(7265):747–753.

McDermott R. (1998). Ethics, epidemiology and the thrifty gene: biological determinism as a health hazard. *Soc Sci Med.* 47:1189–1195.

Min C, Yu Z, Kirsch KH, et al. (2009). A loss-of-function polymorphism in the propeptide domain of the LOX gene and breast cancer. *Cancer Res.* 69(16):6685–6693.

Mittman IS, Downs K. (2008). Diversity in genetic counseling: past, present and future. *J Genet Counseling.* 17:301–313.

Murray CJ, Kulkarni SC, Michaud C, et al. (2006). Eight Americas: investigating mortality disparities across races, counties, and race counties in the United States. *PLoS Med.* 3(9):e260.

Nanda R, Schumm LP, Cummings S, et al. (2005). Genetic testing in an ethnically diverse cohort of high-risk women: a comparative analysis of *BRCA1* and *BRCA2* mutations in American families of European and African ancestry. *JAMA.* 294(15): 1925–1933.

Olopade OI, Grushko TA, Nanda R, Huo D. (2008). Advances in breast cancer: pathways to personalized medicine. *Clin Cancer Res.* 14(24):7988–7999.

Parise CA, Bauer KR, Brown MM, Caggiano V. (2009). Breast cancer subtypes as defined by the estrogen receptor (ER), progesterone receptor (PR), and the human epidermal growth factor receptor 2 (HER2) among women with invasive breast cancer in California, 1999–2004. *Breast J.* 15(6):593–602.

Peltonen L, McKusick VA. (2001). Genomics and medicine. Dissecting human disease in the postgenomic era. *Science.* 291(5507):1224–1229.

Petrucelli N, Daly MB, Bars Culver JO, Feldman GL. (1998). *BRCA1 and BRCA2* hereditary breast/ovarian cancer. In: *GeneReviews* at GeneTests: Medical Genetics

Information Resource [database online]. Seattle, WA: University of Washington. http://www.ncbi.nlm.nih.gov/bookshelf/br.fcgi?book=gene&part=brca1. Updated June 19, 2007. Accessed April 4, 2010.

Risch N. (2006). Dissecting racial and ethnic differences. *N Engl J Med*. 354(4):408–411.

Risch N, Burchard E, Ziv E, Tang H. (2002). Categorization of humans in biomedical research: genes, race and disease. *Genome Biol*. 3(7). Epub Jul 1 2002.

Roden DM, Altman RB, Benowitz NL, et al.; Pharmacogenetics Research Network. (2006). Pharmacogenomics: challenges and opportunities. *Ann Intern Med*. 145(10): 749–757.

Sabatino SA, Coates RJ, Uhler RJ, Breen N, Tangka F, Shaw KM. (2008). Disparities in mammography use among US women aged 40–64 years, by race, ethnicity, income, and health insurance status, 1993 and 2005. *Med Care*. 46(7):692–700.

Shields AE, Burke W, et al. (2008). The use of available genetic tests among minority-serving physicians in the US. *Genet Med*. 10:404–414.

Simon MS, Petrucelli N. (2009). Hereditary breast and ovarian cancer syndrome: the impact of race on uptake of genetic counseling and testing. *Methods Mol Biol*. 471:487–500.

Turnbaugh PJ, Ley RE, Hamady M, Fraser-Liggett CM, Knight R, Gordon JI. (2007). The human microbiome project. *Nature*. 449:804–810.

USA-India Chamber of Commerce Web site. http://www.usaindiachamber.org/. Accessed April 4, 2010.

Vona-Davis L, Rose DP. (2009). The influence of socioeconomic disparities on breast cancer tumor biology and prognosis: a review. *J Womens Health (Larchmt)*. 18(6): 883–893.

Weitzel JN, Lagos V, Blazer KR, et al. (2005). Prevalence of *BRCA* mutations and founder effect in high-risk Hispanic families. *Cancer Epidemiol Biomarkers Prev*. 14(7):1666–1671.

Williams DR, Mohammed SA, Leavell J, Collins C. (2010). Race, socioeconomic status, and health: complexities, ongoing challenges, and research opportunities. *Ann N Y Acad Sci*. 1186:69–101.

[WHO] World Health Organization Commission on Social Determinants of Health. (2008). *Closing the Gap in a Generation: Health Equity through Action on the Social Determinants of Health*. Geneva: World Health Organizatin. http://whqlibdoc.who.int/hq/2008/WHO_IER_CSDH_08.1_eng.pdf.

Zerhouni EA. (2005). US biomedical research: basic, translational, and clinical sciences. *JAMA*. 294(11):1352–1358.

Commentary on the Outcomes Phase of the Translational Cycle

SARA GOERING, SUZANNE HOLLAND, AND KELLY EDWARDS

The outcomes phase of the translational cycle is the place where genetic innovations are expected to finally pay off. Here, at this fourth phase of the cycle, is where we seek to realize positive health impacts as a result of clinical applications of what started out as bench science. We expect to see clinical applications that actually improve health outcomes for people who have been suffering from disease or illness, and who eagerly await, if not a cure, at least improvement. We also hope that data gleaned from the outcomes stage will help us provide meaningful information about prevention for those who are predisposed to certain genetic diseases, though this goal may not always be realizable.

Normative questions related to outcomes research focus our attention on which outcomes are considered most important, who receives the benefits, and whether the purported benefits are always perceived as such. As an important corollary, we can ask how the needs, concerns, and perspectives of recipients are taken into account when new interventions are developed or outcomes are assessed. As Burke and Press have noted (Chapter 10), genetic research on breast cancer is likely to result in a broader range of outcomes benefits if it is informed by social, epidemiological, and physiological studies of breast cancer disparities.

A full assessment of health outcomes may also bring to light the kinds of ancillary services that are needed to assure that a genetic innovation could achieve desired outcomes. For example, if a state newborn screening program (see Chapter 9) provides only screening test results, without follow-up therapy, or if it recommends a regimen that a parent in an underserved community has no hope of enacting (for economic or other access reasons), then the program might actually result in burden rather than benefit for medically underserved families. We have to be critically reflective, then, about whether we have reduced or unwittingly exacerbated existing health inequalities.

Researchers at the outcomes phase must ask themselves whether they are capturing what is of actual importance to recipients of services, or whether they have been sufficiently imaginative about how to achieve those aims. The answering of such questions would be enhanced through partnership with underserved communities, and it would also benefit from greater diversity in the research team. Widening the research perspective to include partners whose disciplinary lenses (clinical, sociological, philosophical, etc.) might help frame the problem differently, or who might understand outcomes in ways not previously considered, would serve this purpose. This is what some have called "knowledge brokering" (Lomas 2007). Knowledge brokering can help us interrogate the usefulness of outcomes research—the actual impact of the research—on the health of the underserved. For example, breast cancer genetics research has focused on the use of genetic testing to identify women at high risk of breast cancer. Yet we know that *BRCA* testing predicts only 5% to 10% of breast cancer cases, and offers only limited prevention options; this type of genetic testing is irrelevant for a majority of breast cancer sufferers. If we asked how researchers might use outcomes data to help reshape the role of genetics in translational breast cancer research, the focus would likely shift to the role of genetics in understanding breast cancer disparities. But, as Burke and Press (Chapter 10) note, genetic explanations of disparities are of limited value if they do not yield strategies for reducing disparities. How could a research program that incorporated both genetic and environmental factors, including social determinants, yield translational value? What would such a program look like? Asking and answering such questions might send researchers back to a more refined and inclusive shaping of the research questions, instead of rushing to market with applications of limited clinical utility that will not necessarily result in improved health outcomes, and which almost certainly won't benefit the underserved.

A risk of focusing on one set of issues—breast cancer genetic testing, for example—is that we may be ignoring the health needs of a larger set of sufferers. For example, are women at risk for breast cancer actually getting mammograms? Are mammograms available and affordable for at-risk women? These questions do not, of course, preclude a role for genetic research, particularly research that sheds light on disease mechanisms and on how social disadvantage may be embodied in cancer disparities. Ideally, the outcomes stage will help us reframe the agenda for the next round of bench research, so that it will be more inclusive of those who had not been recognized as needing to be consulted or included in study designs in the first place. How researchers understand what kind of problem research needs to address, and how they frame that problem, depends in large measure on who is invited to the table and what outcomes matter.

As Riley and Watts discuss in Chapter 9, sometimes health applications are the result of good science *and* the persistent voices of advocates clamoring on behalf of the sufferers, nudging motivated researchers to "find a cure." From the perspective of responsive justice, this is where justice as responsibility insists that when we evaluate the efficacy of health outcomes, we include as partners those communities who suffer, from either disenfranchisement or neglect of recognition in the first place. However, advocacy, as Riley and Watts have shown, is a double-edged sword, for it can also influence the uptake of new discoveries and their applications in the absence of clear evidence of benefit. Expanded newborn screening programs using tandem mass spectrometry, for example, have resulted in an increase in false-positives, and not all newborn screening test results are accompanied by evidence of clinical utility. It is therefore crucial that we interrogate ourselves as researchers, continually asking whom these health outcomes benefit and, sometimes just as importantly, who will likely not benefit, and why not. Such questions naturally lead to a refocusing of priorities, and this highlights the crucial importance of the assessment and priority setting step. Chapter 11 illustrates the kinds of questions and issues that arise at this next step in the translational cycle.

REFERENCE

Lomas J. (2007). The in-between world of knowledge brokering. *Br Med J*. 334:129–132.

11

Bringing the "Best Science" to Bear on Youth Suicide: Why Community Perspectives Matter

ROSALINA JAMES AND HELENE STARKS

When a young person dies by suicide, he or she leaves a legacy of sadness for the family and the community. The unrelenting grief and guilt that can follow such a loss are made worse by all the "if only" thoughts about what might have been done to prevent it. Yet in many—far too many—cases, the tragic legacy of a young person's suicide also signals greater worries for the community. Time and again, what follows is a cascade of attempted and completed suicides that leaves entire communities devastated. American Indian and Alaska Native communities are especially hard-hit, as the headlines show: March 2005: Red Lake Reservation in Minnesota experienced a string of suicides following the terrible killings of nine individuals by a student at Red Lake High School (Indianz.com 2005). January to March 2007: in South Dakota, Rosebud Sioux tribal officials declared a state of emergency after five suicides and more than 100 attempts (Nieves 2007). January to September 2009: 11 young people took their own lives and 51 others attempted suicide on the Standing Rock reservation, which straddles the border of North and South Dakota (Eckroth 2009). December 2008: six deaths and at least as many thwarted attempts occurred in less than three weeks across five communities in northwestern Alaska, triggered by the suicide of a 17-year-old girl from Selawik (Halpin and Hopkins 2008). These are but a few instances of an endemic health disparity with a long history in American Indian and Alaska Native communities.

In 2009, in the aftermath of the youth suicides in her state, Alaska Senator Lisa Murkowski urged the U.S. Department of Health and Human Services to fund a study proposed by Dr. Warren Zapol of the U.S. Arctic Research Commission (USARC) (Murkowski 2009) (see press release, Box 11-1). One of the goals of this research was to reduce the rate of youth suicide in Alaska Native communities by identifying genetic causes that might contribute to depression and substance abuse. The proposed study was to be conducted by the Institute of Medicine (IOM), an independent organization charged with providing decision makers and

---

Box 11-1

TEXT OF PRESS RELEASE FROM SENATOR MURKOWSKI ON ADDRESSING NATIVE YOUTH SUICIDE

Wednesday May 27 2009

WASHINGTON, D.C. – U.S. Sen. Lisa Murkowski, R-Alaska, today asked Health and Human Services Secretary Kathleen Sebelius to fund a $1.2 million study designed to reduce the high rate of suicides among Alaska natives.

In a letter to the Secretary, Murkowski pointed to a study proposed by Commissioner Warren Zapol of the U.S. Arctic Research Commission that would examine the mental and behavioral health issues facing Alaska Native populations living in the Arctic. Zapol's study would be conducted by the Institute of Medicine at the National Academy of Sciences.

This new study would seek to determine the specific genes that contribute to major depressive disorders and alcohol abuse leading to targeted treatment options for Alaska Natives.

"According to the Indian Health Service, suicide rates for American Indians and Alaska Natives are 70 percent higher than the general United States Population," Murkowski wrote. "Suicide is the second leading cause of death for American Indian and Alaska Native youth ages 10-24. Males are especially at risk and commit suicide at a rate five times higher than females."

Murkowski's letter comes in the wake of a May 16, 2009, article in the Anchorage Daily News that highlighted the tragic story of a 17-year-old girl from the community of Selawik, in the Northwest Arctic region of Alaska. The girl's suicide triggered a string of suicide attempts within the community.

"Youth suicides devastate rural communities which are characterized by tightly knit extended families" Murkowski wrote. "The Maniilaq Association, the regional tribal health care provider for the Northwest Arctic, has reported that in 2008, for every suicide in their region, there were 13 suicide attempts."

Murkowski added: "Suicide affects our Native communities in epidemic proportions and we must do all that we can to support our clinicians, communities and leaders to address the issue of youth suicide" she wrote in her letter to Sebelius. "We must bring the best science to Alaska Native communities."

Source: U.S. Senator Lisa Murkowski for the State of Alaska Web site. (2009). Murkowski calls on HHS to address Native youth suicide. http://murkowski.senate.gov/public/index.cfm?p=PressReleases&ContentRecord_id=8315B620-A7DC-02C4-3371-19B14C210C31. Updated May 27, 2009.

---

the public with unbiased advice and recommendations about health issues. Its reports, which are prepared by panels of subject-area experts, are meant to be exhaustive and definitive; they typically take one to two years to complete (IOM 2010). Many IOM studies are mandated by Congress, and study results are often used to set the national research agenda and direct federal funding decisions. Thus, the proposed study had the potential to bring attention and needed resources to this issue of major concern to Senator Murkowski's constituents.

In this chapter, we explore why the IOM study, and the Senator's call for funding, focused so specifically on *genetic* research as a means of addressing the problem of youth suicide in Alaska Native communities. We ask how it is that genetics wins headlines and political capital around this issue, despite substantial evidence from numerous studies, public hearings, and demonstration projects that point to the social determinants of Native youth suicide as a key area for intervention. Returning to the translational cycle set forth in the introduction of this volume, we will use this example to show how the process of setting new priorities in the translational cycle could transform health research practices in the United States.

STATE OF THE SCIENCE: WHAT DO WE KNOW ABOUT THE PROBLEM?

The extent of this particular problem has been described in detail by research reports since the 1950s that have consistently documented epidemic rates of suicide among American Indian and Alaska Native youth, with rates two to four times higher than those seen in European Americans. During this same period, many federal, tribal, and nongovernmental organizations have sponsored research to understand causal factors and determine which intervention strategies are most effective in reaching at-risk youth. A comprehensive review published in 2001 (Middlebrook et al. 2001) and congressional hearings held in 2005 (Committee on Indian Affairs, U.S. Senate 2005) summarize the emerging patterns of risk and protective factors for suicide among American Indian and Alaska Native young people.

Risk factors for youth suicide in Alaska Native communities, such as the Selawik case referred to in Box 11-1, must be understood in light of the history of colonialism, which undermined many very successful and adaptive ways of living in the Arctic. The traumas of colonization include forced removal from traditional lands as well as cultural genocide, which was enacted through sending Native children to residential schools far from their homes and families, prohibition of spiritual ceremonies, banning the use of first languages, and the establishment of child welfare systems that divided families and disrupted the intergenerational transfer of traditional teachings (Morgan and Freeman 2009; Tester and McNicoll 2004; Wexler 2009). Many experts explain the high rates of un(der)employment, poverty, substance abuse, and domestic violence in Alaska Native communities as individual and community responses to acculturation stress related to social, cultural, and economic disruptions (AIPC 2006; Morgan and Freeman 2009; Wexler 2006). In this context, self-destructive behaviors such as substance abuse are understood as coping mechanisms in response to grief and loss of cultural continuity (Chandler and Proulx 2006; Wexler and Goodwin 2006). Young people are especially vulnerable to the consequences of these social dynamics in at least two ways: First, they are forming their identities as they try to walk in two worlds—traditional and modern—with fragile ties and grounding in each. Second, the high incidence of self-destructive behaviors among adults in their communities causes many young people to doubt their own prospects for the future.

This body of research has also identified factors that reduce the risk of youth suicide. A variety of successful interventions have been conducted through school-based curricula, community projects, and culturally relevant clinical environments and treatment centers. A common thread among effective programs is connecting at-risk youth with cultural and spiritual practices that contribute to their moral development and help build a culturally grounded sense of identity and self-esteem. An important conclusion from this research is that overall risk for youth suicide depends on the balance of how families and communities experience both risk and protective factors together. While suicide rates are very high for Alaska Native communities as a whole when compared with other populations, there are significant differences between individual communities. Some experience episodic clusters of completed and attempted suicides within and across family units, while others have no cases at all for decades. In the latter communities, protective factors typically outweigh risk factors.

Running in parallel to the social science research has been more than 30 years of research examining the genetic contributions to suicide risk. Many studies, involving twin and nontwin siblings in biological and nonbiological families, have sought to elucidate the relationship between genes and suicide. These studies consistently demonstrate that when an individual attempts or completes suicide, additional members of his or her family are at elevated risk for suicidal behavior, compared to those with no such family history (Brent and Mann 2005). Research to understand the genetic contribution to suicide risk has focused on genes that potentially influence neurotransmitter activities, particularly on serotonin transport genes and their role in coping with stress (Currier and Mann 2008; Souery et al. 2003). Discussions in this literature focus on how variations in these candidate genes (such as differences in stretches of their genetic sequences) may be related to suicide outcomes, with attention to the role of factors that create and transmit stressful environments (Brodsky et al. 2008; Currier and Mann 2008; Wasserman et al. 2007). Causal links between genes and suicide are thought to be indirect, such that other factors—including patterns of impulsive behaviors and expressions of anger and hostility, especially among males and in the presence of substance use—influence whether a genetic susceptibility will be realized.

What these findings mean for Alaska Native populations is not clear, as most of this research has been done in populations of European descent. Alaska Native communities share a great deal of common ancestry, which suggests that there could be a genetic contribution to higher suicide rates. However, there are wide variations in the suicide rates between individual communities with common ancestry. For example, some villages experience periodic clusters of suicides while others have had periods as great as 30 years without a single suicide. These patterns suggest that other factors besides genetics are also in play.

NEXT STEPS: WHAT DO WE NEED TO KNOW, AND HOW SHOULD WE FIND OUT?

The biomedical and social science communities hold divergent views about how research into the causes and prevention of Alaska Native youth suicide should proceed. Table 11-1 summarizes the different recommendations for future research based on the relevant scientific literatures and our assessment of where these recommendations lead us in the translational cycle. Recommendations for future studies in the biomedical sciences generally focus on further discovery research along two general lines of inquiry: (1) additional investigations into the mechanisms of specific neurotransmitter genes in stress responses, particularly those related to anger, hostility, and impulsivity; and (2) examinations of gene–environment interactions that increase or mitigate the risks for suicide (Brodsky et al. 2008; Currier and Mann 2008; Wasserman et al. 2007). Future research recommendations from the social sciences are summarized in a 2002 IOM report, which calls for surveillance activities to increase the quality and quantity of data collection; community-based interventions that focus on delivery, specifically, activities to disseminate and adapt successful programs to other communities; and evaluation programs to assess the outcomes of interventions aimed at reducing suicide across all communities (Goldsmith et al. 2002).

The lack of overlap between the biomedical and social science recommendations reflects a common problem: all too often, researchers, policy makers, practitioners, and communities are unaware of important, relevant advances in other disciplines. Research that is driven by discipline-specific inquiry seeks to understand the component parts, not the whole, of a problem. This "silo effect"—which keeps valuable knowledge of one domain from informing research and advancing translation in other areas—is especially damaging in health disparities research, given the inherently multifactorial nature of these problems. Working together has the potential to be viewed as compromising the integrity of a disciplinary practice when it requires modifying methodologies and challenging norms to complement the standards of another field (i.e., variations in data collection, statistical analysis, and inclusion of qualitative or mixed methods). However, coordination across disciplines creates opportunities to define the gaps that exist at intersections of these knowledge systems and to better understand the nature of interwoven phenomena, such as the effects of climate change on human physiology and disease patterns. In addition, when research is focused on advancing the body of knowledge in a given field, effecting social change is often less relevant to the research process or the products research delivers. Esteemed values of critical

Table 11-1. SUGGESTIONS FOR THE DIRECTION OF FUTURE RESEARCH

Recommendations for the next phase of the translational cycle

Review articles on the genetics of research	**Discovery**	**Development**	**Delivery**	**Outcomes**
More research on mechanisms of serotonergic genes, genes involved in the stress-response systems (including noradrenaline, the regulation of the stress-related hypothalamic-pituitary-adrenal axis), and genes regulating the speed of neural conduction	X			
Examine gene–environment interactions regarding such factors as the relationship between the above-mentioned genes and the effects of childhood stress (i.e., from sexual abuse, substance use) on development and how it influences adult responses to current stress, mood disorders, aggressive/impulsive traits, and suicidal behavior	X			
Institute of Medicine report: *Reducing Suicide: A National Imperative*	Discovery	Development	Delivery	Outcomes
The National Institute of Mental Health (NIMH), along with other agencies, should develop and support the necessary infrastructure for a national network of suicide research "population laboratories" devoted to interdisciplinary research on safe, high-quality, large-sample, multisite studies on suicide and suicide prevention across the life cycle		X	X	

(continued)

Table 11-1. Suggestions for the Direction of Future Research (continued)

National monitoring of suicide and suicidality through the Centers for Disease Control and Prevention (CDC) and a surveillance system such as the National Violent Death Reporting System that includes data on mortality from suicide. Surveillance activities and all large and/or long-term studies of health behaviors, mental health interventions, and genetic studies of mental disorder should include measures of suicidality (i.e., attempts and completions).	X	X	
Suicide-risk-assessment tools for recognition and screening of patients should be developed and disseminated for primary care clinicians and integrated into medical and nursing school curricula, continuing education, and professional society recommendations for practice	X	X	
Programs for suicide prevention should be developed, tested, expanded, and implemented through funding from appropriate agencies including NIMH, U.S. Department of Veterans Affairs, CDC, and the Substance Abuse and Mental Health Services Administration.		X	X

and objective observation held by the scientific culture are challenged when the role of researchers appears to bridge with advocacy for a health cause. No single discipline, agency, private enterprise, government, or community can independently fulfill the overarching purpose of a research initiative: to create knowledge that has sustainable impact on public health.

An integrated Assessment and Priority Setting step allows stakeholders to create a shared responsibility for maximizing the positive contributions of research in society by connecting research advances to team members that can adopt, modify, or otherwise mobilize higher levels of utility for knowledge. Including a broad range of stakeholders, community advocates, policy makers, diverse academic disciplines, political leadership, and others in reflection and decision making around translational research efforts is more than an enactment of justice. A collaborative model is needed to avoid redundancy in our pursuit of knowledge, move the best basic science or community innovations to intervention, and begin to evaluate the impact of research in real-world settings. Collaboration among stakeholders is required to assure that research effectively informs practice in a way that reduces increasingly complex health inequities. As a group, stakeholders can keep one another accountable for enabling hand-offs along the cycle, and they can ensure that research doesn't stall due to limitations and motivations of individual members.

We now return to Senator Murkowski's press release and ask why genetics research was specifically called out as an important focus for research to "reduce" high rates of Native youth suicide, when comprehensive inquiries have consistently pointed to interdisciplinary and social approaches as key points of intervention. Reexamining the scope of the problem through the lens of genetics and science may well lead to new insights, but the question remains: will these new insights change the outcomes for Alaska Native youth contemplating suicide? It seems particularly salient to examine the optimism surrounding highly technical endeavors in this case, where the pressing need for solutions in real time was stressed. We suggest that two pervasive frames—"medicalization" (construing social problems as medical ones that can, and should, be addressed through the health care system) and "geneticization" (positioning genetics as the best way of understanding and solving medical problems)—have become the taken-for-granted ways of thinking about the causes of and responses to population health problems in the United States. These frames have a powerful way of capturing attention and resources and are constantly reinforced through the popular media and the scientific literature. With respect to the problem of Native youth suicide, these two frames often promote genetics as an optimal solution, overshadowing the equally (or arguably more) compelling evidence from interdisciplinary social and behavioral initiatives.

The National Institutes of Health (NIH) invested more than $4 billion in funded research in the period between 2003 and 2006. This remarkable public investment in human genome and genetics discovery reflects a predominant value in U.S. culture of embracing exploration and supporting ventures into unknown territory (Pohlhaus and Cook-Deegan 2008). Genetics became the new frontier when specific gene mutations were found to cause severe inherited diseases such as cystic fibrosis, in 1989 (Cystic Fibrosis Foundation 2010), and Huntington disease, in 1993 (Hogarth 2003). The media and public continually respond to new findings with enthusiasm, and genomic innovation seems to hold exciting potential for identifying our inherited flaws and addressing them through personalized medicine and pharmacogenomics. Thus framing health issues in terms of genetics can bring attention and cachet, and possibly increased research funding, to a given problem, but it can also have unintended negative effects on the translational cycle.

In Chapter 2 of this volume, our colleague Patricia Kuszler laid a foundation for understanding the social, political, and economic forces that incentivize the investment of research resources toward the development of health products such as medical procedures and new drugs. In the ongoing competition for limited research funds, genetic research is often presented as an important way to address a myriad of health-related issues, including complex diseases such as cancer, diabetes, and autism spectrum disorders. The underlying storyline can be summarized as follows: (1) there is a genetic basis for the disease; (2) we will find the gene(s) that cause it; and (3) when we do, discoveries and products that result from this investment will help people with the disease (Conrad 2001). New genomics tools may indeed hold promise for improving targeted prevention and treatment, but this storyline gives priority to biomedical and pharmaceutical innovations—that is, the kind of expensive new products that are often unavailable to medically underserved populations. Groups that have limited resources and difficulty accessing even basic health services are unlikely to benefit from such innovations.

Breast cancer screening among Native American and Alaska Native women provides a telling example. Although the Centers for Disease Control funds a program for breast cancer screening in some of these communities, the service often cannot be delivered. Mobile mammography units are dependent on good weather conditions to reach remote villages that are accessible only by air or sea. When ferry crossings are cancelled due to bad weather, the cancer screening visits must also be cancelled. For women living in these remote places, convenient access to free mammography is a much higher priority than genetic testing for breast cancer susceptibility (to identify known cancer-causing mutations in the BRCA genes), genetic counseling services, and follow-up care for BRCA+ women.

Framing health issues in terms of medicine and genetics focuses attention on the individual (or genetically related subpopulations) as the point of intervention. When specific genetic risk factors are identified, the thinking goes, pharmacologic and other individualized therapies can be tailored to address those particular factors. We see evidence of this thinking in the press release (Box 11-1), which states that the proposed research study "would seek to determine the specific genes that contribute to major depressive disorders and alcohol abuse leading to targeted treatment options for Alaska Natives."

Explanations of youth suicide that are framed in terms of biomedical pathways of disease that lie within the individual, tend to divert attention and resources away from the social, economic, and political conditions that affect the overall health of Alaska Native people. Focusing on an individual's genetic makeup makes it easier to ignore that person's social position, cultural environment, and opportunities to change his or her living conditions, all of which are known to affect gene expression (Lippman 1991). When health disparities research looks for genetic explanations and interventions without considering these factors as well, it risks increasing racial and ethnic differences in health status and health care (Bonham et al. 2009).

There are several key consequences to underserved populations of allowing genetics to continue to be a primary focus of research efforts. One is opportunity cost: when research resources are dedicated to genetics, they cannot be invested elsewhere. It is unclear how much time and how many research dollars must be invested in pursuit of novel approaches before clinical or public health benefit can be delivered (Hudson 2008). However, in the face of endemic issues such as youth suicide, we must consider the cost of diverting resources from well-defined factors known to promote physical and mental wellness in individuals regardless of the gene variants (Chaufan 2007). Likewise, when priority-setting becomes bogged down in the long-standing debate about nature versus nurture—whether we are the product of our biology, our environment, or a combination of the two—the urgency to relieve the burdens of suicide can be lost. In practicing responsive justice (see Chapter 1), we must address these challenges or risk delaying or failing to achieve the goal of reducing suicide rates (Arnason and Hjörleifsson 2007).

Another result of the focus on genetic pathways to suicide is that initiatives aimed at making tangible changes in near-term outcomes must either compete for funding with technically novel proposals that capture the imagination of decision makers, or allocate resources to incorporate a genetics component into their proposal. Adding genetic discovery activities to projects (i.e., to investigate gene–environment interactions) may make them more competitive, but it may also commit investments to several intermediate steps before translation is complete. This can have the unintended consequence of postponing short-term benefits for the target population. Such delays can have very real costs to those seeking actionable response to the urgency of a suicide epidemic in Alaska Native villages: existing behavioral research, treatment programs, and local community approaches that are ready to move further down the translational pathway (i.e., programs that have reached the development or delivery phases) may be underfunded or remain stagnant.

There is a long record of research into strategies to address the alarming rates of suicide, and these investigations have prompted numerous legislative initiatives. However, many of the proposed changes have yet to be realized. In 1999, the Surgeon General issued a call to action on this same topic which produced a great deal of basic research and recommendations for delivery and outcomes interventions (Table 11-1) that could be used to improve mental health systems and implement locally adapted youth suicide interventions. These initiatives have contributed to overall improvements of services and health care delivery over the past decade, but their impact on suicide and violent-injury rates has been minimal. The resources devoted to these particular issues have been relatively minimal compared to the magnitude of need in the population, especially in light of rising medical costs, chronically underfunded behavioral health programs, and a growing Alaska Native population (Alaska Native Health Board 2009). Justice requires action: affected communities must not be made to wait any longer for this acquired knowledge to be put to work in helping change the outcomes of youth suicide. When investments in past national efforts have yet to be realized, perhaps the translational process can be mobilized by asking why recommendations that could have a positive impact on this chronic issue have not been taken up.

TRANSFORMING RESEARCH THROUGH INCLUSION AND REFLECTION

While the pathway by which research is "translated" into health outcomes has traditionally been understood as a linear process, the conceptual model proposed in this volume envisions a continuous cycle in which health outcomes research and discovery science are connected and inform each other. The transition point between these phases often goes unremarked upon; we argue that there is a need to explicitly attend to this pivotal moment between the current state of affairs and the focus for the next round of investments in research. This critical time focuses on the central phase of the translational cycle and the work of Assessment and Priority Setting, which recognizes the values and choices—both implicit and explicit—that drive how health problems are defined and how resources are allocated. Decisions made at this point in the process of translation can reinforce social structures that maintain the status quo and tend to disenfranchise vulnerable groups, or they can challenge those structures, focus on the needs of vulnerable groups, and prioritize the development of sustainable interventions to address those needs. The work done in Assessment and Priority Setting should reflect commitments to responsive justice (Goering, Holland, and Fryer-Edwards 2008) by assuring that the values and assumptions of all the stakeholders in translation efforts are represented and heard in priority setting, and that there is broad participation in determining what counts as evidence and in deciding the direction and focus of future research.

Outcomes researchers describe three necessary conditions for an intervention to be effective: (1) a health care system that can deliver it; (2) clinicians or public health practitioners who adopt it as practice; and (3) access for members of the target population. We expand the notion of access to include a fourth prerequisite: the target population must be in a position to adopt and implement the intervention into their current social, cultural, and economic environments.

To achieve this fourth element, members of the target population must be included throughout the translational cycle as partners in the research process, not simply as end users of its results and products. The participation of community members in the process is critical, not least because they have expert knowledge that is essential when identifying potential barriers to achieving health benefit from health research, as well as in finding creative and feasible solutions. For example, clinical mental health services in many rural Alaska communities are often provided by non–Native Alaskan clinicians who visit once or twice a month (Wexler and Goodwin 2006). Given this limited access to clinical resources, successful suicide prevention must go beyond surveillance of at-risk youth and referral to licensed psychologists. Prevention work must include participation by community members in developing interventions that match the range of cultural, social, and health resources available to them on an on-going basis.

The involvement of community members in planning, implementing, and evaluating interventions also challenges the methods, definitions of rigor, and disciplinary boundaries of science to move toward collaborative models. For example, the *Elluam Tungiinun* ("Toward Wellness" in the Yup'ik language) prevention program was codeveloped by researchers at the Center for Alaska Native Health Research at the University of Alaska Fairbanks in partnership with Yup'ik village elders, parents, and youth. The goal of the program was to identify individual, family, and community-level protective and risk factors for alcohol use and suicide. Community involvement was vital not only for defining elements of the program, but also for its implementation and sustainability (Allen et al. 2009). Including these critical stakeholders increases the potential for interdisciplinary collaboration and for extending the reach of research beyond the scope of laboratories, academic knowledge banks, and clinical applications toward community adoption and practice.

We return again to the Alaska example to illustrate how integrating an inclusive, reflective phase can improve the translational process, ensure that appropriate goals are selected, and speed up the time required for end users to realize the benefits of research. Leading up to Senator Murkowski's call for research into the genetic causes of depression and alcohol abuse was a multiyear effort spearheaded by USARC Chair Mead Treadwell and Dr. Zapol, also a USARC commissioner. USARC was created under federal law in 1984 to establish goals and policy regarding research plans for the Arctic, including research regarding natural resources and materials; work in the physical, biomedical, and health sciences; and studies in the social and behavioral sciences. In 2007, USARC made it a priority to expand its programs to increase human capacity for research in the Arctic, including engagement in partnership with indigenous communities to enhance their ability to contribute to, and benefit from, research and science (USARC 2007).

Commissioners Treadwell and Zapol spent several years working with various government agencies to raise awareness of the Native youth suicide issue and gain support for research funding. Despite their efforts, they received an underwhelming response from the 27 institutes and centers of the NIH responsible for biomedical research. They then redirected their energies toward convening a meeting of key Alaska stakeholders to obtain their perspectives on Native youth suicide and to garner support for the proposed IOM study. The USARC commissioners enlisted help from the John E. Fogarty International Center for Advanced Study in the Health Sciences, a component of the NIH that addresses global health challenges through international partnerships. The Fogarty International Center ultimately organized and funded the Behavioral and Mental Health Research in the Arctic Strategy Setting Meeting, held in Anchorage the first week of June 2009, just a few days after Senator Murkowski's press release was issued (Levintova, Zapol, and Engmann 2010).

We highlight this meeting as an example of an activity that brought key stakeholders together to foster development of a collective research agenda. There were five stated goals for the two-day meeting: (1) to inventory prior work done on mental health and suicide for Alaska Natives; (2) to learn what works; (3) to examine ways to increase the rigor of interventions and associated outcomes evaluations; (4) to set a course for scaling up successful interventions; and (5) to secure long-term funding to sustain those interventions until the problem of suicide is adequately addressed. These goals focus on delivery and outcomes issues, but the meeting itself represents a process of mobilizing major interest groups to reinvigorate the translation process.

Participants in the meeting (N = 63) included representatives from Alaska Native organizations (n = 14; 22%), academic institutions from Alaska and other states (n = 26; 41%), federal funding agencies (n = 13; 21%), social and clinical services providers (n = 7; 11%) and the Alaska state government (n = 3; 5%). The proceedings from the meeting have been published as a supplement to the *International Journal of Circumpolar Health* (Levintova, Zapol, and Engmann 2010). From our review of the meeting notes, it appears that there was true dialogue and participation among these diverse stakeholders in the scientific presentations and subsequent discussions. Although the majority of the scientific presentations were given by academics, all stakeholders were represented, and all attendees participated in sessions in which the evidence for evaluating research needs, data needs, and next steps was discussed.

Key themes from the meeting reflect and reinforce the principles of responsive justice and the importance of broad participation in decision making about next steps in the translational cycle. These included recognition of the ways in which severe historic traumas have led to severe survival responses, which in turn have distorted interpersonal relationships and cultural practices that previously maintained balance with the social and physical environments. Participants from all the stakeholder groups reported the need for and value of community involvement in designing and implementing research and intervention programs. They also endorsed the importance of community participation in maintaining a focus on models that build on community strengths, such as uniting youth and elders in activities that reinforce cultural identity and create a sense of hope and purpose for youth. Academic researchers emphasized the benefits they derived personally from including community members as partners and teachers in their research, such as learning traditional practices that promote healthy lifestyles, in addition to the positive impact of inclusion on intervention outcomes.

There was a good deal of discussion about the need for a mix of research methods and the value of both statewide epidemiological data and village-level investigations in understanding between-community variation. When medically underserved groups like Alaska Native communities are small in number, they are frequently misrepresented or misunderstood in the context of large population studies. Participants advocated for the inclusion of qualitative, community-based approaches to supplement population studies and elucidate hidden details or processes that contribute to the success or failure of interventions.

A commitment to reflective and inclusive activities like this meeting comes with its own costs for the translational cycle, which also must be considered. It takes considerable time and expense to develop trust and meaningful partnerships between community and academic stakeholders, especially in places such as remote, rural Alaska communities. Trust is established through the practice of face-to-face gatherings, following up on promised actions, and respecting the need for short-term as well as long-term benefits to all the partners involved in the research. Other costs include the time it takes for investments in building capacity for both community members and researchers to pay off. While these investments may take years to show benefit, they create the infrastructure that facilitates translation and prevents the process from stalling due to ignorance of other stakeholders' work or lack of opportunities for appropriate handoffs.

For investigators, research institutes, and funding agencies, another real cost includes sharing power and control over the translational and research processes with stakeholder partners. Processes must be devised that respect differences in how people conceptualize problems and solutions, and in how they value approaches to addressing these. Building interdisciplinary teams that incorporate community participation requires researchers to demonstrate patience, humility, and a willingness to engage other ways of knowing over the methods and perspectives that come from investing in a particular disciplinary worldview. A collaborative dialogue inclusive of diverse skill sets and perspectives reduces communication barriers that can arise when well-intentioned stakeholders are challenged to examine their assumptions about causal pathways of disease and illness, as well as about appropriate interventions. The Assessment and Priority Setting step is more than a process of consultation with token stakeholder representatives; it is a means to develop a common goal that provides benefits for everyone involved. This requires a transformation of all stakeholders toward mindfulness of the power that dominant contemporary frames (such as medicalization and geneticization) can have on how and where problems and solutions are located and defined.

In the short term, there may be different winners and losers when decision making is shared and the appropriation of resources shifts from one discipline to another, or from academic institutions to community-level interventions, to fund different activities across the translational cycle. Investments made in discovery science may preclude the possibility of directing funding toward later phases of translation (i.e., developing, adapting, or sustaining interventions in health and social services systems), and vice versa. Funders will have to accept the costs of this more inclusive approach to agenda setting as a legitimate part of the research process. This will mean increasing active outreach efforts to include community partners in all phases of research planning and review—for example, shaping the direction of long-term strategic objectives and scientific review of proposals.

It will also require accepting indigenous knowledge and approaches as equally valid and rigorous methods for designing and implementing research, and for defining what constitutes evidence of the success or failure of the research endeavor.

CONCLUSION

Dr. Elias Zerhouni, the former director of the NIH, led the call for interdisciplinary research practices as the way to more effective translational research (Zerhouni 2003, 2005). In the case of youth suicide, achieving health benefit in affected communities requires that we move beyond status quo assumptions to incorporate values of responsive justice by (1) redistributing power over the research process to include those closest to understanding the multiple dimensions of suicide in these communities; (2) integrating existing disciplinary, clinical, and Alaska Native indigenous knowledge into translational research hypotheses to ensure adoption and sustainability of downstream deliverables; and (3) delivering meaningful and timely results and products (i.e., interventions, resources, and services) to the groups that the research dollars were intended to help. After 60 years of documenting the problem of youth suicide in Alaska Native communities, it is time to work across disciplines and ensure translation of culturally relevant, deliverable, and sustainable interventions. Members from these communities must be considered equal partners in translational research and share control of the process, especially when research agendas are being set and funding decisions are being made. Their expert knowledge of the resources and ongoing changes in their environment must be included at each phase to anchor new knowledge development of whatever type (discovery science, intervention development, delivery, or outcomes) in the context of the challenges and strengths of their communities, and to assess how this new knowledge will make a difference in reducing rates of Native youth suicide. We agree with Senator Murkowski that the "best science" must be brought to bear on youth suicide; what we argue here is that what counts as the "best science" should be open to interpretation by a coordinated dialogue that incorporates multiple disciplinary and stakeholder perspectives.

REFERENCES

[AIPC] Alaska Injury Prevention Center, Critical Illness and Trauma Foundation, American Association of Suicidology. (2006). *Alaska Suicide Follow-back Study Final Report*. www.hss.state.ak.us/suicideprevention/pdfs_sspc/sspcfollowback2-07.pdf.

Alaska Native Health Board. (2009). *Alaska Native Health Board State Legislative Priorities for Fiscal Year 2009*. https://www.alaskatribalhealth.org/caucus/upload/ANHB-State.pdf.

Allen J, Mohatt G, Fok CC, Henry D; People Awakening Team. (2009). Suicide prevention as a community development process: understanding circumpolar youth suicide prevention through community level outcomes. *Int J Circumpolar Health.* 68(3): 274–291.

Arnason V, Hjörleifsson S. (2007). Geneticization and bioethics: advancing debate and research. *Med Health Care Philos.* 10(4):417–431.

Bonham VL, Citrin T, Modell SM, Hamilton Franklin T, Bleicher EWB, Fleck LM. (2009). Community-based dialogue: engaging communities of color in the United States' genetics policy conversation. *J Health Polit Policy Law.* 34(3):325–359.

Brent DA, Mann JJ. (2005). Family genetic studies, suicide, and suicidal behavior. *Am J Med Genet Part C Semin Med Genet.* 133C(1):13–24.

Brodsky BS, Mann JJ, Stanley B, et al. (2008). Familial transmission of suicidal behavior: factors mediating the relationship between childhood abuse and offspring suicide attempts. *J Clin Psychiatry.* 69(4):584–596.

Chandler M, Proulx T. (2006). Changing selves in changing worlds: youth suicide on the fault-lines of colliding cultures. *Arch Suicide Res.* 10(2):125–140.

Chaufan C. (2007). How much can a large population study on genes, environments, their interactions and common diseases contribute to the health of the American people? *Soc Sci Med.* 65(8):1730–1741.

Committee on Indian Affairs, U.S. Senate. (2005). *Youth Suicide Prevention: Oversight Hearing on the Concerns of Teen Suicide Among American Indian Youths.* Washington, DC: Government Printing Office.

Conrad P. (2001). Genetic optimism: framing genes and mental illness in the news. *Cult Med Psychiatry.* 25(2):225–247.

Currier D, Mann JJ. (2008). Stress, genes and the biology of suicidal behavior. *Psychiatr Clin North Am.* 31(2):247–269.

Cystic Fibrosis Foundation. (2010). About the Cystic Fibrosis Foundation. Cystic Fibrosis Foundation Web site. http://www.cff.org/aboutCFFoundation/PressRoom/AbouttheFoundation/. Accessed December 10, 2010

Eckroth L. (2009, September 29). Standing Rock loses suicide grant. *The Bismarck Tribune.* http://www.bismarcktribune.com/news/state-and-regional/article_5305b16e-ad45-11de-82a6-001cc4c03286.html.

Goering S, Holland S, Fryer-Edwards K. (2008). Transforming genetic research practices with marginalized communities: a case for responsive justice. *Hastings Cent Rep.* 38(2):43–53.

Goldsmith SK, Pellmar TC, Kleinman AC, Bunney WE, eds; Committee on Pathophysiology and Prevention of Adolescent and Adult Suicide, Board on Neuroscience and Behavioral Health. (2002). *Reducing Suicide: A National Imperative.* Washington, DC: National Academies Press.

Halpin J, Hopkins K. (2008, December 19). Rash of teen suicides rocks Northwest Alaska. *Anchorage Daily News.* http://www.adn.com/2008/12/19/627621/rash-of-teen-suicides-rocks-northwest.html.

Hogarth P. (2003). Huntington's disease: a decade beyond gene discovery. *Curr Neurol Neurosci Rep.* 3(4):279–284.

Hudson K. (2008). The health benefits of genomics: out with the old, in with the new. *Health Aff (Millwood).* 27(6):1612–1615.

Indianz.com. (2005). Deadly tragedy puts focus on Native youth problems. http://64.38.12.138/News/2005/007217.asp. Updated March 24, 2005.

[IOM] Institute of Medicine. About the IOM. http://www.iom.edu/About-IOM.aspx. Updated October 12, 2010.

Levintova M, Zapol WI, Engmann E. (2010). *Behavioral and Mental Health Research in the Arctic: Strategy Setting Meeting*. Circumpolar Health Supplements. (5).

Lippman A. (1991). Prenatal genetic testing and screening: constructing needs and reinforcing inequities. *Am J Law Med*. 17(1-2):15-50.

Middlebrook DL, LeMaster PL, Beals J, Novins DK, Manson SM. (2001). Suicide prevention in American Indian and Alaska Native communities: a critical review of programs. *Suicide Life Threat Behav*. 31(Suppl):132–149.

Morgan R, Freeman L. (2009). The healing of our people: substance abuse and historical trauma. *Subst Use Misuse*. 44(1):84–98.

[Murkowski] U.S. Senator Lisa Murkowski for the State of Alaska Web site. (2009). Murkowski calls on HHS to address Native youth suicide. http://murkowski.senate.gov/public/index.cfm?p=PressReleases&ContentRecord_id=8315B620-A7DC-02C4-3371-19B14C210C31. Updated May 27, 2009.

Nieves E. (2007, June 9). Indian reservation reeling in wave of youth suicides and attempts. *New York Times*. http://www.nytimes.com/2007/06/09/us/09suicide.html.

Pohlhaus JR, Cook-Deegan RM. (2008). Genomics research: world survey of public funding. *BMC Genomics*. 9:472.

Souery D, Oswald P, Linkowski P, Mendlewicz J. (2003). Molecular genetics in the analysis of suicide. *Ann Med*. 35(3):191–196.

Tester FJ, McNicoll P. (2004). Isumagijaksaq: mindful of the state: social constructions of Inuit suicide. *Soc Sci Med*. 58:2625–2636.

[USARC] U.S. Arctic Research Commission. (2007). *Report on Goals and Objectives for Arctic Research 2007 for the U.S. Arctic Research Plan*. Arlington, VA and Anchorage, AK: U.S. Arctic Research Commission.

Wasserman D, Geijer T, Sokolowsi M, Rozanov V, Wasserman J. (2007). Nature and nurture in suicidal behavior, the role of genetics: some novel findings concerning personality traits and neural conduction. *Physiol Behav*. 92(1-2):245–249.

Wexler L. (2009). Identifying colonial discourses in Inupiat young people's narratives as a way to understand the no future of Inupiat youth suicide. *Am Indian Alsk Native Ment Health Res*. 16(1):1–24.

Wexler L, Goodwin B. (2006). Youth and adult community member beliefs about Inupiat youth suicide and its prevention. *Int J Circumpolar Health*. 65(5):448–458.

Wexler LM. (2006). Inupiat youth suicide and culture loss: changing community conversations for prevention. *Soc Sci Med*. 63(11):2938–2948.

Zerhouni E. (2003). Medicine. The NIH roadmap. *Science*. 302(5642):63–72.

Zerhouni EA. (2005). Translational and clinical science—time for a new vision. *N Engl J Med*. 353(15):1621–1623.

12

Conclusion

KELLY EDWARDS, SARA GOERING,
SUZANNE HOLLAND, AND MAUREEN KELLEY

"If there is no transformation inside each of us, all the structural change in the world will have no impact on our institutions." (Block 1993)

GENOMIC MEDICINE AND HEALTH DISPARITIES

This book began with a premise about the importance of addressing the needs of the marginalized and medically underserved, and asked how advances in genomic medicine might affect existing health disparities in the United States. Our primary concern has been that a push for greater efficiency of translation of genomic science—from discovery research into development, and on to delivery and outcomes—without attention to matters of health equity in benefits at each phase may exacerbate rather than ameliorate such health disparities. As a consequence, we have argued for a new model of the translational cycle—one that highlights the importance of critical assessment and priority setting between every phase of research—and active engagement by researchers and the research enterprise with marginalized and underserved populations throughout the cycle.

As the push for genomic medicine gains traction and available funds are used to scale up novel genetic testing rather than offering possibly more cost-effective but still not fully implemented preventive care (see Chapter 10 on *BRCA* testing), or as the funding of genetic research is prioritized over other, perhaps more effective social interventions (see Chapter 11 on youth suicide prevention), it is important to make these implicit value trade-offs explicit in public debate, informed by considerations of fairness and principles of equity. Furthermore, even if basic and cost-effective care were available to all, disparities might also increase if effective genomic innovations could only be afforded by those with the ability to pay out of pocket, as suggested by what has been called the "inverse equity hypothesis" (Victora et al. 2000). This hypothesis has been posited to explain data showing that new technologies tend to be adopted only by the wealthy and, therefore, often widen the gap between the wealthy and those living in poverty. In addition, it is possible that disparities could also expand even where access to effective genomic innovations is held constant, because of existing inequities in environmental circumstances. For instance, a clinically valid genetic test for increased susceptibility to cardiovascular disease due to exposure to air pollution might affect people in different environmental situations quite differently. The wealthier at-risk person could move to an area with a lower concentration of air pollution, whereas the impoverished individual might have no such option. Accordingly, the health disparity between individuals in such groups might widen as a result of the differential ability to respond to the information provided by the test. The more research emphasizes genetic factors in health and disease, the more readily people will be tempted to locate the problems inside individuals (and hence affix blame or responsibility there) rather than in the environments or social structures in which they live.

The problems raised above may appear to be primarily about differential access to health care, genomic innovation, and healthy environments. However, as we have seen throughout this book, access to medical technologies and other goods may not be the only thing we need to advance health, particularly if the technologies have been conceived, developed, and implemented into clinical practice without close consultation with the communities for which they are intended. Instead, discussions about access must be accompanied by recognition of marginalized and underserved groups, and engagement with their perspectives, throughout the research cycle. We argue that reconceptualizing *how* the work of genetic research gets done will be important if we are to make any progress on closing the disparities gap. Transformed research practices will both improve the science and produce better, more just results.

TRANSFORMING RESEARCH PRACTICES

We have seen throughout the book that even small changes in the way science is conducted can make a difference by moving research toward discoveries that will promote health for all, rather than widen gaps between the haves and have-nots. Bringing on collaborators from other disciplines—for example, public health, anthropology, or health sciences—could help a project address immediate health needs or preventive health dimensions to a problem, even while longer-term genetic insights are explored. Evidence increasingly shows that involvement of interdisciplinary collaborators (team science) and partnerships with communities can positively impact every dimension of scientific practice—from the nature of the research questions asked, to the type of data collected, to the interpretation and analysis of the findings, and on to the application and effective dissemination of the results (Jordan, Gust, and Scheman 2005; Wallerstein and Duran 2006).

Involving new collaborators helps frame scientific investigation in a way that is more likely to positively impact health outcomes, because the data collected are more relevant, or the health intervention designed will actually be taken up and used by the target population.

Initiating partnerships with target communities, as in community-based participatory research, is one way that research practices can be transformed to aim for more translational science goals and outcomes. However, not all genetic researchers will be able to partner with diverse others due to the constraints of funding structures, institutional culture, or other disciplinary pressures. In these instances, we maintain that the individual researcher still has a responsibility to direct his or her research toward translational goals. Such researchers can meet their responsibilities by attending to, and facilitating, a handoff to researchers or developers at the next phase of research. The usual end point for scientific research—a peer-reviewed publication—is not enough to ensure that the author's findings are going to be taken up and used in the next phase of research or practice. The recurring assessment and priority setting step within the translational research cycle highlights the importance of researchers conferring with others to assess how to move their research findings forward.

Asking researchers at every phase to focus on health impacts, particularly for the most underserved, need not eliminate every opportunity to carry out so-called pure science. Indeed, we acknowledge and appreciate that it can be difficult to predict scientific discoveries, or to know in advance how any discovery may be best used. No doubt serendipity plays a significant role, and will continue to do so. However, broadening any individual researcher's scope of what might be relevant could open the door to new insights. There are increasing examples of interdisciplinary or community-based researchers who have an "aha" moment because of a clinical challenge or an empirical observation. For example, one genetic researcher routinely traveled out to a community for site visits and, in the course of exchanging greetings, noticed that the residents had warm hands despite below-freezing temperatures. This observation led him to pursue a previously unexplored hypothesis about a metabolic function that had a protective effect over diabetes (Boyer oral communication, October 2009). Bench researchers can take such insights back to the lab for further discovery-based work.

Engagement with marginalized communities can offer both epistemic and ethical advantages. Partnering with communities and considering their needs throughout the translational cycle will serve to improve the chances of combating persistent health disparities, because the scientific results will be relevant to the communities. As philosopher of science Naomi Scheman writes, "Communities are the repositories of the narratives in terms of which the researchers' findings will—for better or worse—be—or fail to be—integrated into the wild; and the multiplicity and contradictoriness of those narratives are reason for, not against, engaging with them. Unlike the logic of the laboratory, based on abstraction and generalization, community-based knowledge is based on logics of salience and connection, of particularity and idiosyncracy, on specificities of space and time, on history and hope" (Scheman 2008). If science is to be not only analytically valid but also practically meaningful, researchers will have to grapple with such complications.

We acknowledge that it is not a simple task to try and direct one's research toward a goal of health benefits for all. Determining which benefits to pursue, and who counts them as benefits, is a challenge. Which groups deserve priority and who speaks for any group are additional difficult and unavoidable questions. For instance, when researchers aim to address the needs of people with autism, they may be torn between listening to parents of children with autism, often parents who are desperate for a cure (as with the parents who helped to start Autism Genetic Resource Exchange [AGRE], described in Chapter 4), or listening to autistic children and adults, who may need assistance to communicate but who sometimes tell a very different story about the desirability of a cure. We argue that health impacts should be conceived in a broad manner, to include not only interventions on the medicalized body, but also strategies to improve biopsychosocial well-being more generally.

The foregoing examples demonstrate that transformed research practices are possible, but *why* would a genetic researcher be motivated to target his or her work toward health benefits for all? One reason is that the nature of public funding in the National Institutes of Health (NIH)-focused research enterprise may heighten incentives to focus on research that will be likely to pay off for the public good. Public goods can arguably come in many forms, including increased economic development, scientific and technological capacity building, or enhanced scientific literacy, that may have nothing to do (at least directly) with improved health outcomes. These are all laudable benefits. However, increasingly the public is interested in holding public institutions accountable for promises made. NIH purports to support advances in science that will lead to improvements in health. NIH-funded researchers must strive to make good on that promise, or risk losing public trust in the whole research enterprise.

In addition to investigator-level choices that can be made about partnerships or handoffs, the lessons from this book also point to significant systems-level changes that would help facilitate just research with positive health impacts. As illustrated throughout the book, the current incentive structure for advancing research developments privileges drugs, tests, and devices that will corner a market. What alternative, or counterbalancing, incentives would need to be in place to advance discoveries with a greater public health impact? Funders could start requiring more detailed dissemination plans as part of a grant proposal, including guidance and review criteria for effective handoffs or collaborations. Academic credit is currently heavily weighted toward peer-reviewed publications and grant dollars secured. Instead of mainly tracking impact factors of journal publications, authors and researchers could be given credit for partnerships sustained, outcomes impacted, or public communication/education produced (see www.ccph.info for such proposals; CCPH 2010).

Genetics is a highly dynamic field and changes are already afoot, even as this book goes to press. By its nature, our model of responsive justice is iterative and, as such, is well suited to move along with the field. As new directions or initiatives in science emerge, researchers can run through a series of questions to assess whether the proposed direction meets concerns about justice and utility. Individual researchers can make their own assessments and choices, even while larger-scale policy and social changes are under consideration or in development.

TAKING RESPONSIBILITY

While our book urges a fairly significant transformation in the way that translational science is undertaken and understood, initiating such changes need not be daunting for the individual scientist. A small but valuable first step would be to simply acknowledge the difficult position of the marginalized and underserved, and to consider how one might contribute to the struggle for justice within one's area of scientific expertise. For example, scientists within the field of public health genetics and maternal health are turning their attention to deeply puzzling gene–environment interactions that may contribute to pregnancy loss or premature birth; such determinants include workplace risks or household pollutants, common risks for women living at or beneath the poverty line (Stillerman et al. 2008; Windham and Fenster 2008) Seeing the problems of the marginalized and underserved as significant and pressing is an important first step. Arguing for a similar point about engagement with communities, Scheman (2008, p. 122) comments: "Scientific researchers, even those most comfortable in laboratories, have important roles to play, but they can play those roles well—even by epistemic standards—only insofar as they think of themselves and the knowledge they create as framed by, and responsible to, the relationships in which, whether they recognize it or not, they are enmeshed." This book is a call to make those relationships explicit, and to encourage scientists at every phase of the translational cycle to attend to their significance.

LIMITATIONS AND EXPANSIONS

There are certainly limitations to the perspectives and examples we have laid forth in this volume. Foremost among them is the fact that the examples and discussion are directed exclusively to the research policies and practices in the United States. The focus on *genetic* research may also be seen as too narrow for some, particularly in light of the need for team science to pull us toward discoveries that will have a likely health impact. Our task was to engage in problems that have arisen in the U.S. context of genetic research; however, the examples addressed here have significant parallels elsewhere. The responsive-justice model can provide a useful framework for thinking about equity and benefit in respect to global health as well as within other areas of scientific research.

For example, with growing global awareness about disparities in disease burden, a new challenge will be to improve translational applications of discovery science to low- and middle-income countries where the disease burden is greatest (Black et al. 2009; London and Kimmelman 2008). Many populations outside the United States face significantly greater barriers to good health than those faced by even the worst-off Americans. Individuals in such populations need access to medical care, to improvements in the social determinants of health, and to emerging technologies that may improve health. As importantly, though, they need to be recognized as essential partners in addressing their health issues. Indeed, the obligation to recognize and address diseases shouldered by the worst-off globally is increasingly being seen not as a duty of rescue but as an obligation to improve investment in research capacity and investigators in these regions and countries. Furthermore, partnering with such regional investigators will bring the knowledge of local customs, living conditions, and political histories that are necessary for ensuring that available health interventions are designed to have the greatest opportunity for success. The responsibility element of responsive justice pushes researchers to acknowledge and work to address global disparities in health, while the recognition element insists that in so doing, they build research partnerships with local communities in need. Such partnerships will ensure that details about the meaning of the disease burden are not assumed but understood from those with firsthand experience. In addition, it is important to consider equity measures in relation to distribution over the long term and with sensitivity to the inverse-equity hypothesis associated with the introduction of new technologies described above (Victora et al. 2000).

To address concerns about the distribution of benefits, one group of child health investigators has advocated the implementation of equity criteria to assess the impact new interventions have in widening the gap between rich and poor within and between communities (Barros et al. 2009). To illustrate how considerations of equity can be incorporated in a longitudinal study, consider work done on child survival and development in Mali, where researchers focused not only on *reaching* underserved segments of a population (an issue of access), but also on demonstrating how the intervention they offered decreased the relative disease burden on the poorest group within that population (an issue of health outcomes). According to the UNICEF Accelerated Child Survival and Development (ACSD) program in Mali, five years after an ACSD intervention focused on outreach to rural mothers, access to antenatal care was significantly more equitable in districts with ACSD than in other districts, though both the intervention and comparison areas showed marked social gradients before the program was implemented in 2001 (Bryce et al. 2008). The intervention explicitly targeted *equitable* access, not merely access for some. Improving discovery science translation for low- and middle-income countries will require creative thinking for implementing promising systems biology research in these settings, as we are seeing in HIV and TB research, and, more recently, in maternal and child health research (Kelley and Rubens 2010).

CONCLUSION

This brief discussion of global health inequities illustrates how the responsive-justice concerns we have outlined in the U.S. context may have broader applicability and resonance. Our goal is to push what is possible within the work of science by asking critical questions of justice at all phases of the translational cycle, from assessment and priority setting to discovery, development, delivery, and outcomes research.

Throughout this volume we have seen examples of investigators and communities coming together for a common purpose: to advance scientific research to improve health. Our intention in using the responsive-justice framework as a lens through which to examine, question, challenge, and motivate genetic research is to illustrate that different choices can be made throughout the translational research cycle—choices that have a greater likelihood of impacting the health of the public, particularly those who suffer unfairly from health inequities. As the brief global health examples illustrate, responsive justice is an approach that can be utilized in many types of research settings and contexts, each with its own unique adaptation and application.

Rather than focusing on efficiency in translation from discovery research to health applications, we have argued for the importance of an explicit acknowledgment of matters of justice throughout the translational cycle. In our model, such justice matters involve elements of distribution, recognition, and responsibility. Through arguments and examples offered in the chapters and commentaries throughout the book, we have recommended a way forward that will ensure that the needs of individuals and populations suffering significantly from health inequities are kept in view as genetic research progresses. We hope that at least some researchers and teams will be inspired by the ideas advanced in this book, and will initiate their own transformations in response. We recognize, of course, that the power of any individual researcher or team to effect change is constrained by larger structures and traditions that can seem intimidating and unalterable. But improvements in such structures and traditions will only be able to occur, after all, when concerned and committed individuals acknowledge injustice and take the steps to initiate change.

REFERENCES

Barros FC, Victora CG, Scherpbier RW, Gwatkin D. (2009). Inequities in the health and nutrition of children. In: Blas E, Sivasankara Kurup A, eds. *Priority Public Health Conditions: From Learning to Action on Social Determinants of Health.* Geneva: World Health Organization.

Black RE, Bhan MK, Chopra M, Rudan I, Victora CG. (2009). Accelerating the health impact of the Gates Foundation. *Lancet.* 373(9675):1584–1585.

Block P. (1993). *Stewardship: Choosing Service over Self-Interest.* San Francisco, CA: Berrett-Koehler Publishers, Inc.

Bryce J, Gilroy K, Jones G, Hazel E, Black RE, Victora CG. (2008). *Final Report of the Retrospective Evaluation of ACSD: Cross-Site Analyses and Conclusions.* Baltimore, MD: Johns Hopkins Bloomberg School of Public Health, Institute for International Programs.

[CCPH] Campus-Community Partnerships for Health Web site. http://www.ccph.info/. Accessed September 12, 2010.

Jordan C, Gust S, Scheman N. (2005). The trustworthiness of research: the paradigm of community-based participatory research. *Metropolitan Universities*. 16(1):39–57.

Kelley M, Rubens C. (2010). Global report on preterm birth and stillbirth (6 of 7): ethical considerations. *BMC Pregnancy and Childbirth*. 10(Suppl 1):S6.

London AJ, Kimmelman J. (2008). Justice in translation: from bench to bedside in the developing world. *Lancet*. 372(9632):82–85.

Scheman N. (2008). Narrative, complexity and context: autonomy as an epistemic value,. In: Lindemann H, Verkerk M, Walker M., eds. *Naturalized and Normative Bio-Ethics*. Cambridge University Press: 106-124.

Stillerman KP, Mattison DR, Giudice LC, Woodruff TJ. (2008). Environmental exposures and adverse pregnancy outcomes: a review of the science. *Reprod Sci*. 15(7):631–650.

Victora CG, Vaughan JP, Barros FC, Silva AC, Tomasi E. (2000). Explaining trends in inequities: evidence from Brazilian child health studies. *Lancet*. 356(9235): 1093–1098.

Wallerstein NB, Duran B. (2006). Using community-based participatory research to address health disparities. *Health Promotion and Practice*. 7(3):312–323.

Windham G, Fenster L. (2008). Environmental contaminants and pregnancy outcomes. *Fertil Steril*. 89(2Suppl):e111–e116.

INDEX

Note: Page numbers followed by "*f*" and "*t*" denote figures and tables, respectively.